MW00531834

Transactions on Computational Science and Computational Intelligence

Series Editor
Hamid Arabnia
Department of Computer Science
The University of Georgia
Athens, Georgia
USA

Computational Science (CS) and Computational Intelligence (CI) both share the same objective: finding solutions to difficult problems. However, the methods to the solutions are different. The main objective of this book series, "Transactions on Computational Science and Computational Intelligence", is to facilitate increased opportunities for cross-fertilization across CS and CI. This book series will publish monographs, professional books, contributed volumes, and textbooks in Computational Science and Computational Intelligence. Book proposals are solicited for consideration in all topics in CS and CI including, but not limited to, Pattern recognition applications; Machine vision; Brain-machine interface; Embodied robotics; Biometrics; Computational biology; Bioinformatics; Image and signal processing; Information mining and forecasting; Sensor networks; Information processing; Internet and multimedia; DNA computing; Machine learning applications; Multiagent systems applications; Telecommunications; Transportation systems; Intrusion detection and fault diagnosis; Game technologies; Material sciences; Space, weather, climate systems, and global changes; Computational ocean and earth sciences; Combustion system simulation; Computational chemistry and biochemistry; Computational physics; Medical applications; Transportation systems and simulations; Structural engineering; Computational electro-magnetic; Computer graphics and multimedia; Face recognition; Semiconductor technology, electronic circuits, and system design; Dynamic systems; Computational finance; Information mining and applications; Astrophysics; Biometric modeling; Geology and geophysics; Nuclear physics; Computational journalism; Geographical Information Systems (GIS) and remote sensing; Military and defense related applications; Ubiquitous computing; Virtual reality; Agent-based modeling; Computational psychometrics; Affective computing; Computational economics; Computational statistics; and Emerging applications. For further information, please contact Mary James, Senior Editor, Springer, mary.james@springer.com.

More information about this series at http://www.springer.com/series/11769

E. S. Gopi

Pattern Recognition and Computational Intelligence Techniques Using Matlab

E. S. Gopi
Department of Electronics &
Communications Engineering
National Institute of Technology Trichy
Tamil Nadu, India

ISSN 2569-7072 ISSN 2569-7080 (electronic)
Transactions on Computational Science and Computational Intelligence
ISBN 978-3-030-22272-7 ISBN 978-3-030-22273-4 (eBook)
https://doi.org/10.1007/978-3-030-22273-4

© Springer Nature Switzerland AG 2020
This work is subject to copyright. All rights are reserved by the Publisher, whether the whole or part of the material is concerned, specifically the rights of translation, reprinting, reuse of illustrations, recitation, broadcasting, reproduction on microfilms or in any other physical way, and transmission or information storage and retrieval, electronic adaptation, computer software, or by similar or dissimilar methodology now known or hereafter developed.
The use of general descriptive names, registered names, trademarks, service marks, etc. in this publication does not imply, even in the absence of a specific statement, that such names are exempt from the relevant protective laws and regulations and therefore free for general use.
The publisher, the authors, and the editors are safe to assume that the advice and information in this book are believed to be true and accurate at the date of publication. Neither the publisher nor the authors or the editors give a warranty, express or implied, with respect to the material contained herein or for any errors or omissions that may have been made. The publisher remains neutral with regard to jurisdictional claims in published maps and institutional affiliations.

This Springer imprint is published by the registered company Springer Nature Switzerland AG.
The registered company address is: Gewerbestrasse 11, 6330 Cham, Switzerland

Dedicated to my wife, Ms. G. Viji; my son, Mr. A.G. Vasig; and my daughter, Miss. A.G. Desna.

Preface

The book summarizes various dimensionality reduction techniques such as PCA, LDA, KLDA, and ICA.

This also discusses various linear regression techniques such as parametric approach, Bayes technique, and kernel method.

The discriminative approach such as Support Vector Machine, the probabilistic discriminative model such as logistic regression, the probabilistic generative model such as Hidden Markov Model and Gaussian Mixture Model, and the various computational intelligence techniques such as Particle Swarm Optimization, ANT colony technique, and Artificial Neural Network are discussed in this book. The recent popular techniques such as convolution network, autoencoder, and Generative Adversarial Network are summarized. The various statistical tests applicable to pattern recognition techniques are also reported. This book is meant for those who are doing basic and applied research in machine learning, pattern recognition, and computational intelligence.

Tamil Nadu, India E. S. Gopi

Acknowledgments

I would like to thank Prof. Mini Shaji Thomas (Director, NIT, Tiruchirappalli), Prof. S. Soundararajan (Former Director, NIT, Tiruchirappalli), Prof. M. Chidambaram (IITM, Chennai), Prof. K. M. M. Prabhu (IITM, Chennai), Prof. B. Venkataramani (NIT, Tiruchirappalli), and Prof. S. Raghavan (NIT, Tiruchirappalli) for their support. I would also like to thank those who helped me directly or indirectly in bringing out this book successfully. Special thanks to my parents E. Sankara Subbu and E. S. Meena. I also thank the research scholars Ms. G. Jayabrindha, Mr. P. Rajasekhar Reddy, and Ms. K. Vinodha for helping in proofreading the manuscript.

Tamil Nadu, India E. S. Gopi

Contents

Chapter 1
Dimensionality Reduction Techniques

This chapter focuses on various dimensionality reduction techniques and the methodology to achieve Gaussianity of the given data.

1.1 Principal Component Analysis (PCA)

Consider the random vector $\mathbf{u}$ and $\mathbf{v}$. Let the number of elements in the vectors $\mathbf{u}$ and $\mathbf{v}$ are m and n respectively, with $n < m$. The methodology involved in obtaining the transformation matrix $\mathbf{W}$ (with size $m \times n$) that maps the random vector $\mathbf{u}$ to $\mathbf{v}$ as $\mathbf{v} = \mathbf{W}^T u$ is known as dimensionality reduction techniques. Let us explore the dimensionality reduction with $m = 2$ and $n = 1$. Note that in this case, $\mathbf{W}$ is the vector of size 2×1. We need the constraint to obtain the better choice for the matrix $\mathbf{W}$. Let us choose the constraint as $\mathbf{W}^T\mathbf{W} = 1$. Also, let us compute the variance of the transformed random variable V as follows:

$$\sigma_v^2 = E(\mathbf{v}^2) - (E(\mathbf{v}))^2 \tag{1.1}$$

$$= E(\mathbf{W}^T\mathbf{u}\mathbf{u}^T\mathbf{W}) - (E(\mathbf{W}^T\mathbf{u})E(\mathbf{u}^T\mathbf{W})) \tag{1.2}$$

$$= \mathbf{W}^T(E(\mathbf{u}\mathbf{u}^T) - E(\mathbf{u})E(\mathbf{u}^T)\mathbf{W} \tag{1.3}$$

$$= \mathbf{W}^T\mathbf{C_U}\mathbf{W} \tag{1.4}$$

where $\mathbf{C_u}$ is the co-variance matrix of the random vector $\mathbf{u}$. Let us optimize the vector $\mathbf{W}$ that maximizes the variance of the random variable v. Formulating the Lagrangian equation is obtained as follows. By maximizing $\mathbf{W}^T\mathbf{C_U}\mathbf{W}$, subject to the constraint $\mathbf{W}^T\mathbf{W} = 1$.

$$\mathbf{W}^T\mathbf{C_U}\mathbf{W} + \lambda(\mathbf{W}^T\mathbf{W} - 1) \tag{1.5}$$

© Springer Nature Switzerland AG 2020

E. S. Gopi, *Pattern Recognition and Computational Intelligence Techniques Using Matlab*, Transactions on Computational Science and Computational Intelligence,
https://doi.org/10.1007/978-3-030-22273-4_1

Let $\mathbf{W} = [w_1\ w_2]^T$ and $\mathbf{C} = \begin{bmatrix} c_{11} & c_{12} \\ c_{12} & c_{22} \end{bmatrix}$.

Note that the matrix $\mathbf{C_U}$ is symmetric. Equation (1.5) is rewritten as follows:

$$c_{11}w_1^2 + 2c_{12}w_1w_2 + w_2^2c_{22} - \lambda(w_1^2 + w_2^2 - 1) \tag{1.6}$$

Differentiating (1.6) with respect to w_1, w_2, and λ to obtain the following:

$$2c_{11}w_1 + 2c_{12}w_2 - 2\lambda w_1 = 0 \tag{1.7}$$

$$2c_{12}w_1 + 2c_{12}w_2 - 2\lambda w_2 = 0 \tag{1.8}$$

$$\mathbf{CW} = \lambda\mathbf{W} \tag{1.9}$$

In this case, **W** is the eigenvector of the matrix **C**. Thus the eigenvector of the co-variance matrix **C** is in the direction of maximum variance of 2D data. Therefore, the projection of 2D data using the transformation matrix (eigenvector **W**) is the orthogonal projection of 2D data on the eigenvector, which is in the direction of maximum variance of 2D data. This is an example for mapping 2D data to 1D data using Principal Component Analysis (PCA).

1.1.1 Illustration of Projection of 2D to 1D Data

Two class 2D data (with mean vector $[1\ 1]^T$ and $[-1\ -1]^T$) are generated. They are subjected to dimensionality reduction using the transformation matrix obtained using Principal Component Analysis (PCA). It is noted that the vector E_1 is obtained as the eigenvector corresponding to the largest eigenvalue of the co-variance matrix (computed using the generated 2D data). Figure 1.1 illustrates the projection of 2D data on the eigenvector.

pca2d1d.m

```
pca2d1d.m
X=randn(2,100)*0.5+repmat([1 1]',1,100);
Y=randn(2,100)*0.5+repmat([-1 -1]',1,100);
plot(X(1,:),X(2,:),'r*');
hold on
plot(Y(1,:),Y(2,:),'b*');
Z=[X Y];
C=cov(Z',1);
[E D]=eig(C);
hold on
plot([-4*E(1,2) 4*E(1,2)],[-4*E(2,2) 4*E(2,2)],'-')
plot([-4*E(1,1) 4*E(1,1)],[-4*E(2,1) 4*E(2,1)],'-')
```

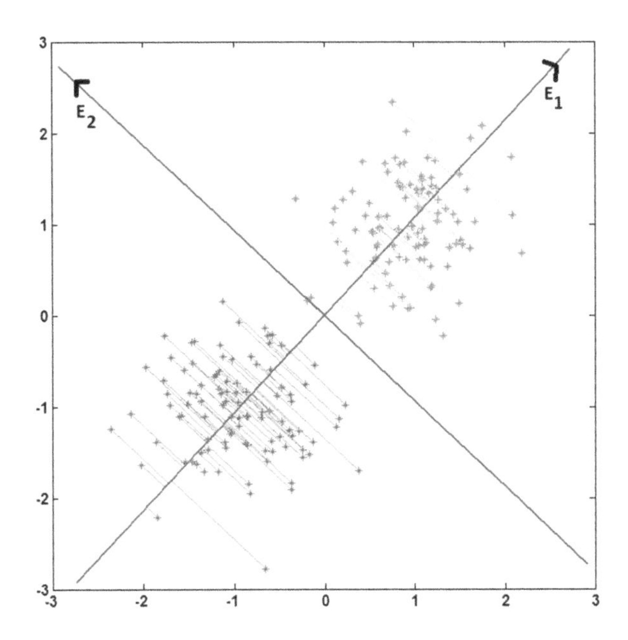

Fig. 1.1 Illustration of the projection of the 2D data on the eigenvector E_1, which is in the direction of maximum variance of the 2D data

```
for i=1:1:100
VECTOR1=E(:,2)'*X(:,i)*E(:,2);
VECTOR2=E(:,2)'*Y(:,i)*E(:,2);
plot([X(1,i) VECTOR1(1)],[ X(2,i) VECTOR1(2)],'c-')
hold on
plot([Y(1,i) VECTOR2(1)],[ Y(2,i) VECTOR2(2)],'m-')
hold on
end
```

1.1.2 PCA

In general, the projection of the data from the m-dimensional space to the n-dimensional space with $n < m$ using PCA is summarized as follows:

1. Compute the co-variance matrix of the given data (m-dimensional space).
2. Compute the eigenvectors corresponding to the significant eigenvalues (say n significant vectors) of the computed co-variance matrix.
3. Arrange the obtained eigenvectors columnwise to obtain the transformation matrix $\mathbf{W}$.
4. Project the arbitrary vector $\mathbf{u}$ in the m-dimensional space to obtain $\mathbf{v}$ in the n-dimensional space using the transformation $\mathbf{v} = \mathbf{W}^T\mathbf{u}$.

1.1.3 Illustration

The original 3D data is represented in (1, 1) subplot of Fig. 1.2. The same data is represented in image format in the subplot (1, 2). The subplot (1, 3) gives the plot image of the distance matrix computed using the original 3D data. $(m, n)^{\text{th}}$ element of the distance matrix gives the distance between the mth vector and the nth vector of the original 3D data. If the shade is dark, it indicates that those vectors are closer to each other in the Euclidean space. Similarly, the bright shade indicates that the vectors are far away from each other. The corresponding 2D projected data using two eigenvectors corresponding to two largest eigenvalues and the 1D projected data corresponding to eigenvector corresponding to the largest eigenvalue are illustrated in the second and third rows of Fig. 1.2. The PCA is applied to the face image dataset. The number of face images in each class is 40, and the number of classes is 10. Each vector is of size 48×48, i.e., 2304. The number of significant eigenvalues selected is 5. Thus each vector of size 2304×1 is mapped to the lower dimensional space with size 5×1. The projected data is illustrated in the subplot (2, 1) of Fig. 1.3. The distance matrix corresponding to the data in the higher dimensional space and the projected data is given in the subplot (1, 2) and (2, 2), respectively. It is seen from the distance matrix image that there is no significant difference between the data in the higher dimensional space and the lower dimensional space in terms of Euclidean distance space.

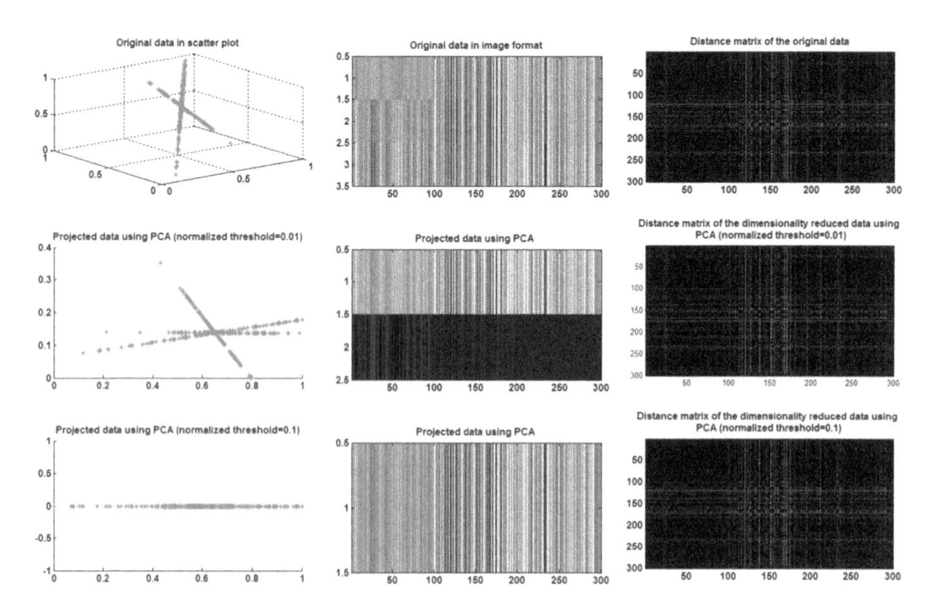

Fig. 1.2 Illustration of Principal Component Analysis (3D data)

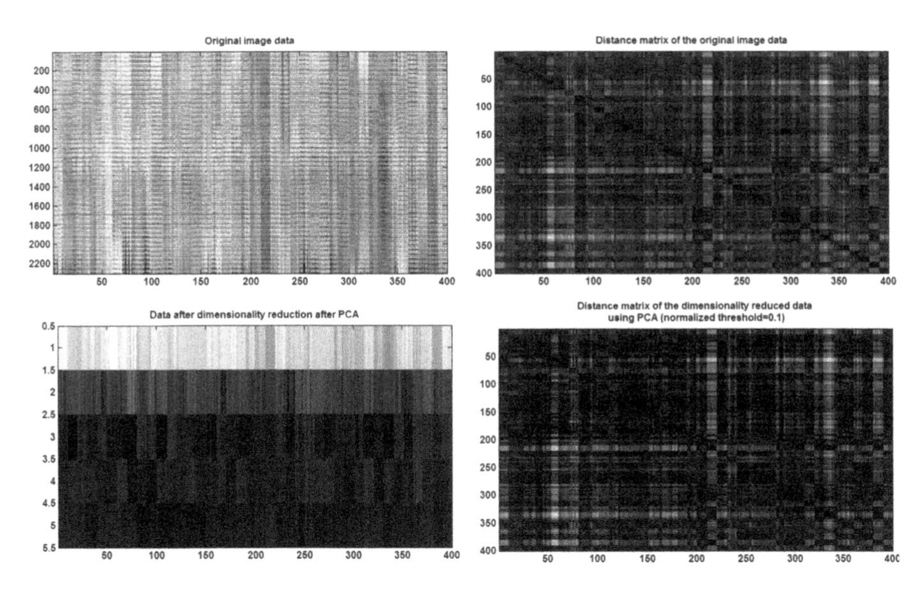

Fig. 1.3 Illustration of Principal Component Analysis (image data). The columns of the image matrix in subplot (1, 1) represent the individual face image data of size 2304×1

pcademo.m

```
%Demonstration of dimensionality reduction techniques

%%%%%%%%%%%%%%%%%%%%%%%%%%%%%%%%%%%%%%%%%%%%%%%%%%%%%
%%%%%%%%%%%%%%%%%%%%%%
%Principl Component Analysis
%With multivariate gaussian data with the specified
mean vector and the specified co-variance matrix
%%%%%%%%%%%%%%%%%%%%%%%%%%%%%%%%%%%%%%%%%%%%%%%%%%%%%
%%%%%%%%%%%%%%%%%%%%%%
m1=rand(3,1)*15-2;m2=rand(3,1)*5+103;m3=rand(3,1)*1+1;
N1=100; N2=100; N3=100;
temp1=rand(length(m1),1)*5-1;C1=temp1*temp1';
temp2=rand(length(m2),1)*5+3;C2=temp2*temp2';
temp3=rand(length(m3),1)*5+1;C3=temp3*temp3';
[X1,C1est]=genrandn(m1,C1,N1);
[X2,C2est]=genrandn(m2,C2,N2);
[X3,C3est]=genrandn(m3,C3,N3);
X=[X1 X2 X3];
X=normalizedata(X);
figure(1)
subplot(3,3,1)
scatter3(X(1,:),X(2,:),X(3,:),'r*')
```

```
title('original data in scatter plot')
subplot(3,3,2)
imagesc(X)
title('original data in image format')
visualizedata(X,3,3,3,1)
title('Distance matrix of the original data')
[E1,Y1,L1]=dimredpca(X,0.01);
Y1=normalizedata(Y1);
subplot(3,3,4)
scatter(Y1(1,:),Y1(2,:),'r*')
title('Projected data using PCA (normalized
threshold=0.01)')
subplot(3,3,5)
imagesc(Y1)
title('Projected data using PCA ')
visualizedata(Y1,3,3,6,1)
title('Distance matrix of the dimensionality
reduced ...
data using PCA (normalized threshold=0.01)')
[E1,Y1,L1]=dimredpca(X,0.1);
Y1=normalizedata(Y1);
subplot(3,3,7)
scatter(Y1(1,:),zeros(1,length(Y1)),'r*')
title('Projected data using PCA (normalized
threshold=0.1)')
subplot(3,3,8)
imagesc(Y1)
title('Projected data using PCA')
visualizedata(Y1,3,3,9,1)
title('Distance matrix of the dimensionality
reduced ...
data using PCA (normalized threshold=0.1)')

%With real world image data
load IMAGEDATA
figure(2)
subplot(2,2,1)
X=normalizedata(X);
imagesc(X)
title('Original image data')
visualizedata(X,2,2,2,2)
title('Distance matrix of the original image data')
[E1,Y1,L1]=dimredpca(X,0.1);
subplot(2,2,3)
Y1=normalizedata(Y1);
```

```
imagesc(Y1);
title('Data after dimensionality reduction after PCA')
visualizedata(Y1,2,2,4,2)
title('Distance matrix of the dimensionality
reduced ...
data using PCA (normalized threshold=0.1)')

figure(7)
visualizedata(X(:,1:1:120),2,2,1,7)
title('Distance matrix of the original image data')
visualizedata(Y1(:,1:1:120),2,2,2,7)
title('Distance matrix of the dimensionality
reduced ...
data using PCA (normalized threshold=0.1)')
```

1.2 Fast Computation of PCA

PCA basis are the eigenvectors of the co-variance matrix computed using the given vectors. Let the co-variance matrix computed using the given column vectors $\mathbf{x_1}, \mathbf{x_2}, ..., \mathbf{x_m}$ of the matrix be $\mathbf{C_x}$.

$$\mathbf{C_x v} = \lambda \mathbf{v} \tag{1.10}$$

The eigenvector $\mathbf{v}$ lies in the space spanned by the vectors used to compute the co-variance matrix $\mathbf{C_x}$. Let the matrix $\mathbf{M} = [\mathbf{x_1}, \mathbf{x_2}, ..., \mathbf{x_m}]$. There exists the vector $\mathbf{u}$ such that

$$\mathbf{v} = \mathbf{Mu} \tag{1.11}$$

This implies the following:

$$\mathbf{C_x v} = \lambda \mathbf{v} \tag{1.12}$$

$$\Rightarrow \mathbf{C_x Mu} = \lambda \mathbf{Mu} \tag{1.13}$$

Multiplying $\mathbf{M^T}$ on both sides, we get the following:

$$\mathbf{M}^T \mathbf{C_x Mu} = \lambda \mathbf{M}^T \mathbf{Mu} \tag{1.14}$$

$$\Rightarrow (\mathbf{M}^T \mathbf{M})^{-1} \mathbf{C_I u} = \lambda \mathbf{u} \tag{1.15}$$

The size of the matrix $\mathbf{C_x}$ is $n \times n$, where n is the number of elements of the column vectors of the matrix $\mathbf{M}$. But the size of the matrix $(\mathbf{M}^T \mathbf{M})^{-1} \mathbf{C_I}$ is $m \times m$, where m is the number of vectors used to compute the co-variance matrix $\mathbf{C_x}$ (refer (1.15)). Also it is observed that the vector $\mathbf{u}$ is the eigenvector of the matrix $(\mathbf{M}^T \mathbf{M})^{-1} \mathbf{C_I}$ with the identical eigenvalue. It is observed that $\mathbf{v}$ is the eigenvector of the matrix $\mathbf{C_x}$ with eigenvalue λ (refer (1.12)). Thus the eigenvector $\mathbf{v}$ is computed

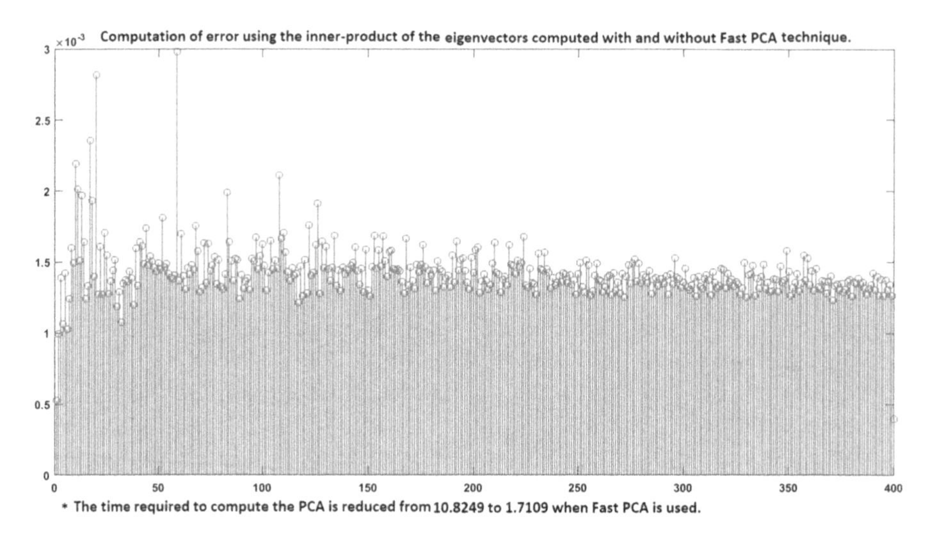

Fig. 1.4 Illustration of the minimal error achieved using fast computation method for PCA basis

using the eigenvector **u**, using **Mu**. The computation time required to compute the eigenvector **u** is much lesser when compared with the computation time required to compute **v**, if $n > m$ (refer Fig. 1.4). Thus, the overall time required to compute eigenvector **v** is reduced by computing u, followed by Mu. The absolute difference between the vector computed with and without Fast PCA algorithm is illustrated in Fig. 1.4.

fastpca.m

```
%Fast computation of PCA using Inner-product matrix
load IMAGEDATA
DATA=double(X)/255;
%Time taken to compute the PCA basis in the direct
method.
tic
M=mean(DATA');
s=0;
for i=1:1:400
    s=s+DATA(:,i)*DATA(:,i)';
end
s=s/400;
s=s-M'*M;
[E1,D1]=eigs(s,400);
time1=toc;
for i=1:1:400
E1(:,i)=E1(:,i)/(sqrt(sum(E1(:,i).^2)));
end
```

```
%Using inner-product technique
%Formulation of Inner-product matrix
tic
s=0;
for i=1:1:400
    for j=1:1:400
        G(i,j)=DATA(:,i)'*DATA(:,j);
    end
end
M=mean(G');
s=0;
for i=1:1:400
    s=s+G(:,i)*G(:,i)';
end
s=s/400;
s=s-M'*M;
[E,D]=eigs(inv(DATA'*DATA)*s,400);
E2=DATA*E;
for i=1:1:400
E2(:,i)=E2(:,i)/(sqrt(sum(E2(:,i).^2)));
end
time2=toc;
figure
plot(sum((E1.*E2).^2))
```

1.3 Linear Discriminant Analysis (LDA)

Let $\mathbf{x_{ij}}$ be the ith vector in the jth class. Let $j = 1, \ldots, K$ and $i = 1, \ldots, n_j$. The between-class scatter matrix is computed as follows:

$$\mathbf{S_B} = \sum_{k=1}^{k=K} n_k(\mathbf{c_k} - c)^T(\mathbf{c_k} - c) \tag{1.16}$$

where $\mathbf{c_k}$ is the centroid of the kth class.

$$\mathbf{c_k} = \frac{1}{K}\sum_{i=1}^{i=K} \mathbf{c_i} \tag{1.17}$$

Similarly, the within-class scatter matrix is computed as follows:

$$\mathbf{S}_B = \sum_{j=1}^{k=K}\sum_{i=1}^{i=n_j} (\mathbf{x_{ij}} - \mathbf{c_j})^T(\mathbf{x_{ij}} - \mathbf{c_j}) \tag{1.18}$$

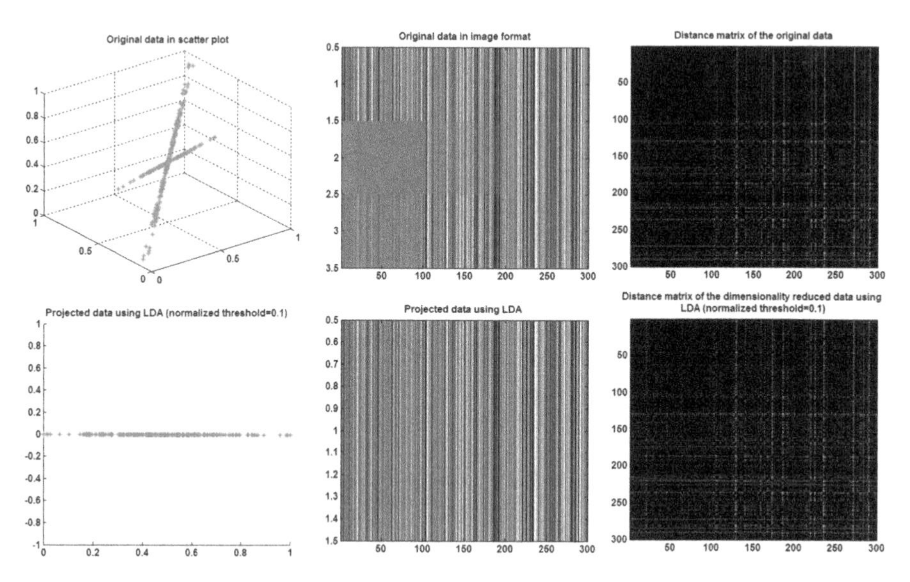

Fig. 1.5 Illustration of Linear Discriminant Analysis (3D data)

The trace of the matrix $\mathbf{S_W}$ measures how the vectors are closer to each other. Similarly, the trace of the matrix $\mathbf{S_B}$ measures how the centroids of various classes are well separated. Let us consider the transformation matrix $\mathbf{W}$ that maps the vector to the lower dimensional space be represented as $\mathbf{y} = \mathbf{Wx}$. The between-class scatter matrix and the within-class scatter matrix computed using the transformed vectors are obtained as $\mathbf{W}S_B\mathbf{W}^T$ and $\mathbf{W}S_W\mathbf{W}^T$, respectively.

The transformation matrix $\mathbf{W}$ is obtained such that $\frac{\text{trace}(\mathbf{W}S_B\mathbf{W}^T)}{\text{trace}(\mathbf{W}S_W\mathbf{W}^T)}$ is maximized. The solution is the eigenvector corresponding to the significant eigenvalues of the matrix $\mathbf{S_W^{-1}S_B}$. The eigenvectors corresponding to the significant values (say r vectors) are arranged row wise to obtain the matrix $\mathbf{W}$ (size $r \times m$), where m and r are the number of elements of the vector before and after projection. This measures how the projected vectors are closer within the class and how the centroids of various classes are mode apart. They are known as LDA basis. The projection of 3D data to the 1D data using LDA basis is illustrated in Fig. 1.5. The illustration of the dimensionality reduction using LDA basis using the image data is given in Fig. 1.6. The distance matrix image obtained using the data in the higher dimensional space and the lower dimensional space (refer Figs. 1.5 and 1.6) reveals that there is no significant loss in the representation of the data in the lower dimensional space.

ldademo.m

```
%%%%%%%%%%%%%%%%%%%%%%%%%%%%%%%%%%%%%%%%%%%%%%%%
%Linear Discriminant Analysis
%%%%%%%%%%%%%%%%%%%%%%%%%%%%%%%%%%%%%%%%%%%%%%%%
m1=rand(3,1)*15-2;m2=rand(3,1)*5+103;m3=rand(3,1)*1+1;
N1=100; N2=100; N3=100;
```

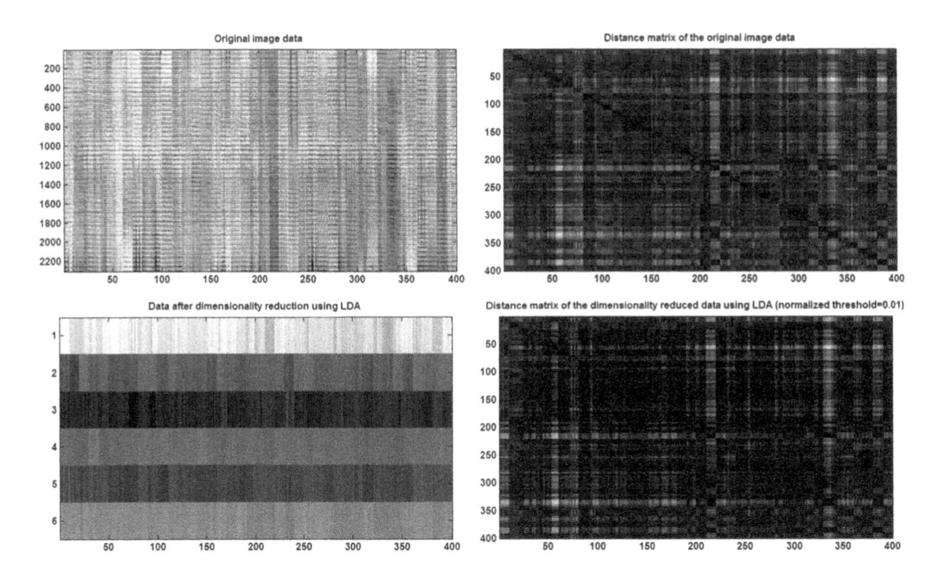

Fig. 1.6 Illustration of Linear Discriminant Analysis (image data)

```
temp1=rand(length(m1),1)*5-1;C1=temp1*temp1';
temp2=rand(length(m2),1)*5+3;C2=temp2*temp2';
temp3=rand(length(m3),1)*5+1;C3=temp3*temp3';
[X1,C1est]=genrandn(m1,C1,N1);
[X2,C2est]=genrandn(m2,C2,N2);
[X3,C3est]=genrandn(m3,C3,N3);
X=[X1 X2 X3];
X=normalizedata(X);
figure(3)
subplot(2,3,1)
scatter3(X(1,:),X(2,:),X(3,:),'r*')
title('original data in scatter plot')
subplot(2,3,2)
imagesc(X)
title('original data in image format')
visualizedata(X,2,3,3,3)
title('Distance matrix of the original data')
[E1,Y1,L1]=dimredlda(X,[100 100 100],0.1);
Y1=normalizedata(Y1);
subplot(2,3,4)
scatter(Y1(1,:),zeros(1,length(Y1)),'r*')
title('Projected data using LDA (normalized threshold
=0.1)')
subplot(2,3,5)
imagesc(Y1)
```

```
title('Projected data using LDA ')
visualizedata(Y1,2,3,6,3)
title('Distance matrix of the dimensionality reduced
data ...
using LDA (normalized threshold=0.1)')

%With real world image data
load IMAGEDATA
figure(4)
subplot(2,2,1)
X=normalizedata(X);
imagesc(X)
title('Original image data')
visualizedata(X,2,2,2,4)
title('Distance matrix of the original image data')
[E1,Y1,L1]=dimredlda(X,ones(1,10)*40,0.1);
Y1=real(Y1);
subplot(2,2,3)
imagesc(Y1);
title('Data after dimensionality reduction after LDA')
visualizedata(real(Y1),2,2,4,4)
title('Distance matrix of the dimensionality reduced
data ...
using LDA (normalized threshold=0.01)')

figure(7)
visualizedata(Y1(:,1:1:120),2,2,3,7)
title('Distance matrix of the dimensionality reduced
data ...
using LDA')
```

1.4 Kernel-Linear Discriminant Analysis

The vectors in the feature dimensional space are mostly not well separated and need to be mapped to the higher dimensional space, where we expect the vectors to be well separated. Thus the vectors are mapped to the higher dimensional space and are brought down to the lower dimensional space using PCA and LDA. These are known as kernel-PCA and kernel-LDA, respectively. Let the vector $\mathbf{x}$ is mapped to the higher dimensional space as $\phi(\mathbf{x})$. The LDA is constructed as follows:

$$\mathbf{S}_{\mathbf{W}}^{\phi} w^{\phi} = \mathbf{S}_{\mathbf{B}}^{\phi} w^{\phi} \tag{1.19}$$

where $\mathbf{S_W^\phi}$ is the within-class scatter matrix computed using the vectors in the higher dimensional space. Similarly, $\mathbf{S_B^\phi}$ is the between-class scatter matrix computed using the vectors in the higher dimensional space. $\mathbf{w}^\phi$ is the LDA basis vector computed in the higher dimensional space. As the eigenvector lies in the space spanned by the vectors used to compute the scatter matrix $\mathbf{S_W^\phi}$ and $\mathbf{S_B^\phi}$, we represent $\mathbf{w}^\phi = \mathbf{M}^\phi \mathbf{u}^\phi$, where the matrix $\mathbf{M}^\phi$ is obtained as follows:

$$\mathbf{M} = [\boldsymbol{\phi}(x_{11})\ \boldsymbol{\phi}(x_{12})\ \boldsymbol{\phi}(x_{13}) \cdots \boldsymbol{\phi}(x_{1n_1})\quad \boldsymbol{\phi}(x_{21})\ \boldsymbol{\phi}(x_{22})$$
$$\boldsymbol{\phi}(x_{23}) \cdots \boldsymbol{\phi}(x_{2n_2}) \cdots \boldsymbol{\phi}(x_{K1})\ \boldsymbol{\phi}(x_{K2})\ \boldsymbol{\phi}(x_{K3}) \cdots \boldsymbol{\phi}(x_{Kn_K})] \quad (1.20)$$

Representing (1.20) as follows:

$$\mathbf{S_W^\phi} \mathbf{w}^\phi = \mathbf{S_B^\phi} \mathbf{w}^\phi \quad (1.21)$$
$$\Rightarrow \mathbf{S_W^\phi} \boldsymbol{\phi} = \mathbf{S_B^\phi} \boldsymbol{\phi} \quad (1.22)$$

Multiplying $\boldsymbol{\phi}^\mathbf{T}$ on both sides of the matrix, we get the following:

$$\boldsymbol{\phi}^\mathbf{T} \mathbf{S_W^\phi} \boldsymbol{\phi} = \boldsymbol{\phi}^\mathbf{T} \mathbf{S_B^\phi} \boldsymbol{\phi} \quad (1.23)$$

It is observed that the matrix $\boldsymbol{\phi}^\mathbf{T} \mathbf{S_W^\phi} \boldsymbol{\phi}$ is the between-class scatter matrix computed in the kernel space. The kernel space is the space spanned by the column space of the matrix $\mathbf{G}$. $(m \times n)^{\text{th}}$ column vector of the matrix $\mathbf{G}$ is given as follows.

$$[k(\mathbf{x_{11}}, \mathbf{x_{mn}})\ k(\mathbf{x_{11}}, \mathbf{x_{mn}}) \cdots k(\mathbf{x_{1n_1}}, \mathbf{x_{mn}}) \cdots k(\mathbf{x_{n_K K}}, \mathbf{x_{mn}})] \quad (1.24)$$

where $k(\mathbf{x}, \mathbf{y})$ is the kernel function. It is noted that the size of the column vector is $n_1 + n_2 + \cdots + n_K$. Thus the size of the matrix $\mathbf{G}$ is given as

$$(n_1 + n_2 + \cdots + n_K) \times (n_1 + n_2 + \cdots + n_K).$$

The eigenvector $\boldsymbol{\phi}$ is the kernel-LDA basis. The column vectors of the transformation matrix ($\mathbf{E}$) are constructed using the kernel-LDA basis (eigenvectors $\boldsymbol{\phi_1}$ $\boldsymbol{\phi_2}$ $\boldsymbol{\phi_3} \cdots \boldsymbol{\phi_r}$, corresponding to the significant eigenvalues) arranged columnwise. Note that the size of the vector $\mathbf{u_i}$ for all i is $(n_1+n_2+\cdots+n_K) \times (n_1+n_2+\cdots+n_K)$. The arbitrary vector $\mathbf{x}$ with size $m \times n$ is mapped to the lower dimensional vector using kernel-LDA described as follows:

1. Map the vector $\mathbf{x}$ to the kernel space as $[k(\mathbf{x_{11}}, \mathbf{x})\ k(\mathbf{x_{11}}, \mathbf{x}) \cdots k(\mathbf{x_{1n_1}}, \mathbf{x}) \cdots \mathbf{k}(\mathbf{x_{n_K} n_K}, \mathbf{x})]$. Let it be $\mathbf{k}(\mathbf{x})$.
2. The obtained vector is mapped to the lower dimensional space using $\mathbf{E^T k(x)}$.
3. Thus the vector in the feature dimensional space with size $m \times 1$ is mapped to the lower dimensional space with size $r \times 1$.

1.4.1 Illustration of K-LDA

Figure 1.7 illustrates the mapping of the vector from 3D space to 1D space. The corresponding distance matrix reveals that there is no significant loss in representing the data in the lower dimensional space (Euclidean space). KLDA is demonstrated using the image dataset. In this case, the image data is projected to 1D space. The zero block matrix seen in the distance matrix reveals that the vectors belonging to the identical class are made closer to each other (Fig. 1.8). The experiments are performed using the Gaussian kernel function (with s^2 as the tuning parameter) as follows:

$$k(x, y) = e^{\frac{(\mathbf{x}-\mathbf{y})^{\mathrm{T}}(\mathbf{x}-\mathbf{y})}{\mathbf{s}^2}} \tag{1.25}$$

kernellda.m

```
%%%%%%%%%%%%%%%%%%%%%%%%%%%%%%%%%%%%%%%%%%%%%%%%%%%%%
%Kernel Linear Discriminant Analysis (Using Gaussian
kernel)
%%%%%%%%%%%%%%%%%%%%%%%%%%%%%%%%%%%%%%%%%%%%%%%%%%%%%
m1=rand(3,1)*15-2;m2=rand(3,1)*5+103;m3=rand(3,1)*1+1;
N1=100; N2=100; N3=100;
temp1=rand(length(m1),1)*5-1;C1=temp1*temp1';
temp2=rand(length(m2),1)*5+3;C2=temp2*temp2';
temp3=rand(length(m3),1)*5+1;C3=temp3*temp3';
[X1,C1est]=genrandn(m1,C1,N1);
[X2,C2est]=genrandn(m2,C2,N2);
```

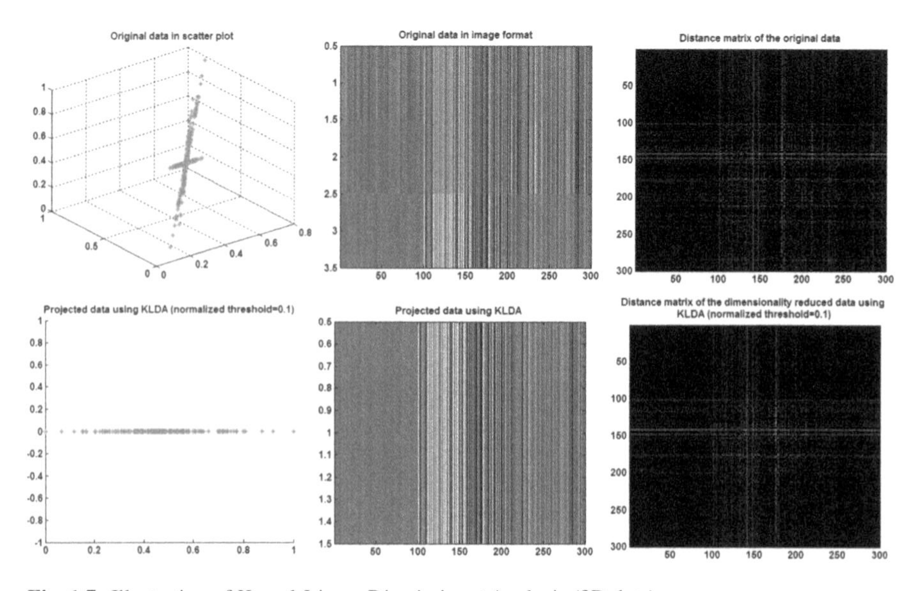

Fig. 1.7 Illustration of Kernel-Linear Discriminant Analysis (3D data)

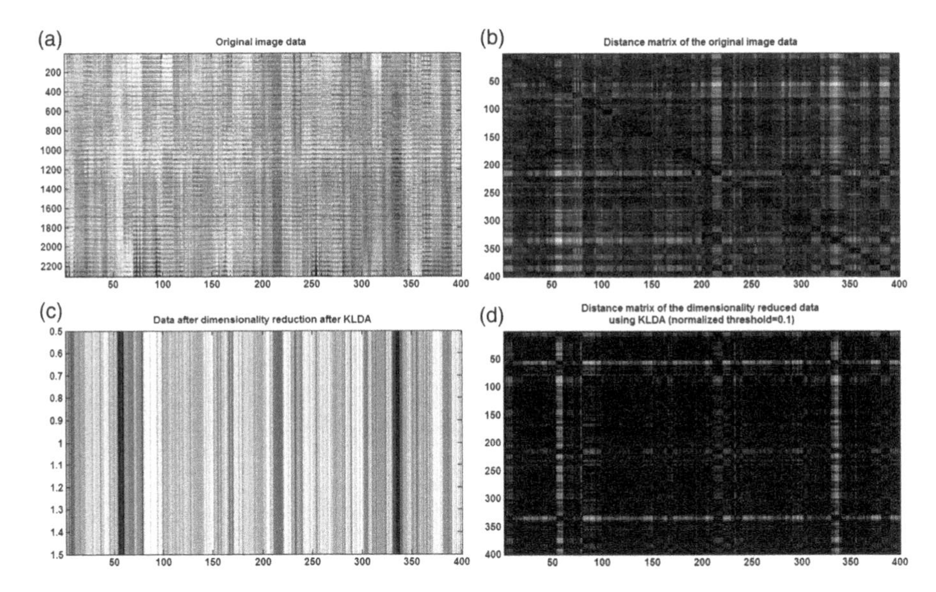

Fig. 1.8 Illustration of Kernel-Linear Discriminant Analysis (image data)

```
[X3,C3est]=genrandn(m3,C3,N3);
X=[X1 X2 X3];
X=normalizedata(X);
figure(5)
subplot(2,3,1)
scatter3(X(1,:),X(2,:),X(3,:),'r*')
title('original data in scatter plot')
subplot(2,3,2)
imagesc(X)
title('original data in image format')
visualizedata(X,2,3,3,5)
title('Distance matrix of the original data')
[E1,Y1,L1]=dimredklda(X,[100 100 100],0.1);
Y1=normalizedata(Y1);
subplot(2,3,4)
scatter(Y1(1,:),zeros(1,length(Y1)),'r*')
title('Projected data using KLDA (normalized
threshold=0.1)')
subplot(2,3,5)
Y1=real(Y1);
imagesc(Y1)
title('Projected data using KLDA ')
visualizedata(Y1,2,3,6,5)
title('Distance matrix of the dimensionality reduced
```

```
data using ...
KLDA (normalized threshold=0.1)')


%With real world image data
load IMAGEDATA
figure(6)
subplot(2,2,1)
X=normalizedata(X);
imagesc(X)
title('Original image data')
visualizedata(X,2,2,2,6)
title('Distance matrix of the original image data')
[E1,Y1,L1]=dimredklda(X,ones(1,10)*40,0.1);
Y1=real(Y1);
subplot(2,2,3)
Y1=normalizedata(Y1);
imagesc(Y1);

title('Data after dimensionality reduction after
KLDA')
visualizedata(real(Y1),2,2,4,6)
title('Distance matrix of the dimensionality reduced
data using ...
KLDA (normalized threshold=0.1)')

figure(7)
visualizedata(Y1(:,1:1:120),2,2,4,7)
title('Distance matrix of the dimensionality reduced
data using KLDA')
```

dimredklda.m

```
function [Esel,Y,L]=dimredklda(X,split,thresh)
%Each column of the variable X represents the data
%Each row of the variable represents the feature
%split represents the number of vectors in the
individual classes.length of
%the split gives the number of classes
%thresh represents the threshold to fix the number of
elements
%in the lower dimensional space
%L is the number of elements in the lower dimensional
space
split=[1 split];
SB=0;
```

```
MEANDATA=[];
for i=1:1:length(split)-1
DATA=X(:,split(i):1:split(i+1));
n=size(DATA,2);
M=mean(DATA');
temp1=(DATA-repmat(M',1,size(DATA,2)));
temp2=temp1*temp1';
SB=SB+n*temp2;
MEANDATA=[MEANDATA M];
end
M=mean(MEANDATA');
temp3=(MEANDATA-repmat(M',1,size(MEANDATA,2)));
SW=temp3*temp3';
SW=real(SW);
[E,V]=eig(pinv(SW)*SB);
V=real(V);
V=diag(V)/max(diag(V));
[P,Q]=find(V>=thresh);
Esel=E(:,size(E,2):-1:P(1));
L=size(Esel,2);
Y=Esel'*X;
```

dimredpca.m

```
function [Esel,Y,L]=dimredpca(X,threshold)
%Each column is representing the data
%Each row is representing the feature
%threshold to fix the number of elements in the lower
dimensional space
%L is the number of elements in the lower dimensional
space
M=mean(X');
temp=(X-repmat(M',1,size(X,2)));
CM=temp*temp';
[E,V]=eig(CM);
V=diag(V)/max(diag(V));
[P,Q]=find(V>=threshold);
Esel=E(:,size(E,2):-1:P(1));
L=size(Esel,2);
Y=Esel'*X;
```

dimredklda.m

```
function [Esel,Y,L]=dimredklda(X,split,thresh)
%Each column of is representing the data
%Each row represents the feature
%split represents the number of vectors in the
```

```
individual classes.length of
%the split gives the number of classes
%thresh represents the threshold to fix the number of
elements
%in the lower dimensional space
%L is the number of elements in the lower dimensional
space
%kf is the kernel function
%Construction of Gram-matrix using Gaussian Kernel
function
for i=1:1:size(X,2)
    for j=1:1:size(X,2)
        temp(i,j)=gaussiankernel(X(:,i),X(:,j),10);
    end
end
X=temp;
SB=0;
MEANDATA=[];
for i=1:1:length(split)
DATA=X(:,1:1:split(1));
n=size(DATA,2);
M=mean(DATA');
temp1=(DATA-repmat(M',1,size(DATA,2)));
temp2=temp1*temp1';
SB=SB+n*temp2;
MEANDATA=[MEANDATA;M];
end
size(MEANDATA)
M=mean(MEANDATA);
temp3=(MEANDATA'-repmat(M',1,size(MEANDATA',2)));
SW=temp3*temp3';
SW=real(SW);
[E,V]=eig(pinv(SW)*SB);
V=diag(V)/max(diag(V))
[P,Q]=find(V>=thresh)
Esel=E(:,1:1:P(1));
L=size(Esel,2);
Y=Esel'*X;
```

visualizedata.m

```
function [Done]=visualizedata(X,p,q,r,s)
%rth subplot (p,q) divided figure s
X=X-min(min(X));
X=X/max(max(X));
D=[];
```

```
for i=1:1:size(X,2)
    for j=1:1:size(X,2)
        D(i,j)=sum((X(:,i)-X(:,j)).^2);
    end
end
D=normalizedata(D);
figure(s)
subplot(p,q,r)
colormap(bone)
imagesc(D);
```

normalizedata.m

```
function [res]=normalizedata(X)
%Makes the input data ranging from 0 to 1
X=X-min(min(X));
res=X/max(max(X));
```

genrandn.m

```
function [RES,C1]=genrandn(m,C,N)
%m is the meanvector (column vector)
%C is the intended co-variance matrix%N is the number
of data
%C1 is the estimate of the co-variance matrix
computed using the generated
%data
DATA=randn(N,length(m));
[E,D]=eig(C);
RES=E*D.^(1/2)*DATA';
RES=real(RES)+m;
%Check the covariance matrix
C1=real(cov(RES'));
```

gaussiankernel.m

```
function [res]=gaussiankernel(x1,x2,sigma)
res=exp(-sum((x1-x2).^2)/(2*sigma^2));
```

1.5 Independent Component Analysis (ICA)

Consider the random variables $x_1, x_2, ..., x_N$ are independent. They are linearly combined to obtain another set of random variables $y_1, y_2, ..., y_N$, using the following linear equation. Let $\mathbf{x} = [x_1\ x_2\ \cdots x_N]^T$ and $\mathbf{y} = [y_1\ y_2 \cdots y_N]^T$ are related as follows:

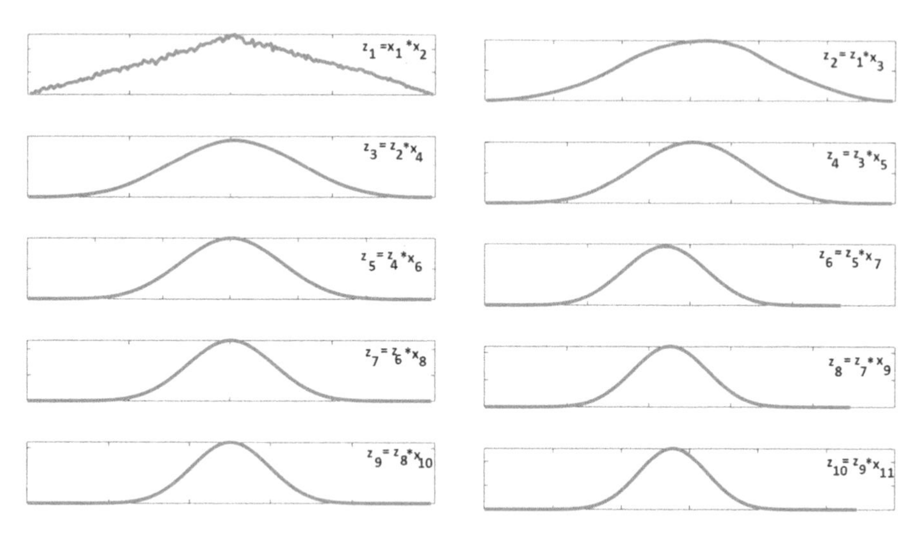

Fig. 1.9 Illustration that summation of uniform distributed random variables is more Gaussian. The subplot (1) is less Gaussian and the subplot (10) is more Gaussian

$$\mathbf{y} = \mathbf{M}\mathbf{x} \tag{1.26}$$

where **M** is the mixing matrix. The requirement is to estimate the mixing matrix **M**, given the outcomes of the random vector **y**. If X_1 and X_2 are random variables that are having the density functions as $f_{X_1}(\alpha)$ and $f_{X_2}(\alpha)$, then the random variable $Z = X_1 + X_2$ follows the density function $f_Z(\alpha) = f_{X_1}(\alpha) * f_{X_2}(\alpha)$.

In particular if X_1 and X_2 are uniformly distributed between 0 and 1, then the density function of Z is triangular in shape. In the same fashion, if $Z = X_1 + X_2 + X_3 + X_4 + \cdots + X_N$, Z tends to follow Gaussian distributed (refer Fig. 1.9). Hence the variables associated with mixtures follow Gaussian distributed, and the random variables associated with the random vector **y** are more Gaussian. Thus the problem is formulated to obtain the matrix **W** such that $\mathbf{W}^T\mathbf{y}$ are more non-Gaussian. The kurtosis of the random variable X with mean 0 is defined as $E[X^4]-3(E[X^2])^2$. The kurtosis value is zero if the random variable X is Gaussian distributed. The kurtosis values of the random variable X (Gaussian) and Y (Gaussian) in Fig. 1.10(a) are minimum, when compared to the absolute kurtosis value of the data mentioned in subplots (b)–(d). Figure 1.11 illustrates the maximization of the absolute values of kurtosis. The matrix **W** is estimated as follows:

1. Apply PCA to obtain the data (random vector **y**) to obtain the random vector **z** that has the co-variance matrix diagonal. By doing so, we are actually making the data uncorrelated.
2. The matrix **W** is estimated such that $\mathbf{W}^T\mathbf{z}$ is less Gaussian, i.e., absolute of the kurtosis value is maximized as described below.
3. (a) Initialize the matrix **W**.

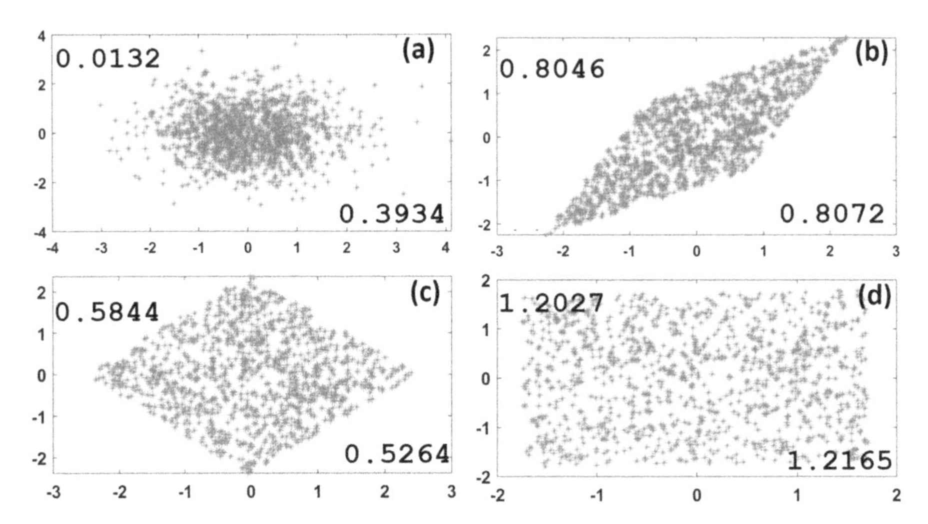

Fig. 1.10 (**a**) Gaussian data. (**b**) Original data. (**c**) After PCA. (**d**) After ICA. (Kurtosis values are mentioned in the subplot.)

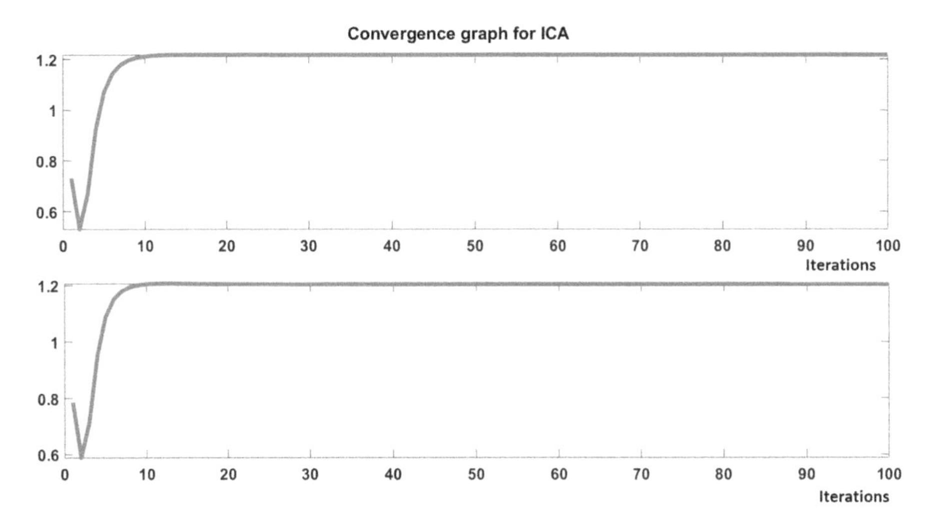

Fig. 1.11 Illustration of the maximization of absolute values of kurtosis

(b) The rows of the matrix $\mathbf{W}$ are made orthogonal using Gram–Schmidt orthogonalization procedure. Suppose the vectors $\mathbf{v_1}$, $\mathbf{v_2}$ and $\mathbf{v_3}$ are independent, then the magnitude of the vector $\mathbf{v_1}$ is made 1 as $\mathbf{q_1} = \frac{\mathbf{v_1}}{||\mathbf{v_1}||}$. The orthogonal vector corresponding to the vector $\mathbf{v_2}$ is obtained as $\mathbf{u_2} = \mathbf{v_2} - (\mathbf{v_2}^T\mathbf{q_1})\mathbf{v_1}$, and the corresponding orthonormal vector is given as $\mathbf{q_2} = \frac{\mathbf{u_2}}{||\mathbf{u_2}||}$. Similarly, the orthonormal vector which is orthogonal to both $\mathbf{v_1}$ and $\mathbf{v_2}$ is given as follows: $\mathbf{u_3} = \mathbf{v_3} - (\mathbf{v_3}^T\mathbf{q_1})\mathbf{q_1} - (\mathbf{v_3}^T\mathbf{q_2})\mathbf{q_2}$ and $\mathbf{q_3} = \frac{\mathbf{u_3}}{||\mathbf{u_3}||}$.

(c) Let the matrix $\mathbf{W}^T = \begin{bmatrix} \mathbf{w_1} \\ \mathbf{w_2} \\ \mathbf{w_3} \end{bmatrix}$ where $\mathbf{w_i}$ is the ith row of the matrix. Let $\mathbf{w_i} = [w_{i1}\ w_{i2}]$. The vector $\mathbf{w_i}$ is multiplied with the random vector $\mathbf{z}$ to obtain the random vector $\mathbf{y}$. The vector $\mathbf{w_i}$ is obtained by maximizing the absolute of the kurtosis (J) as follows:

$$J = abs(E((\mathbf{w_i z})^4) - 3(E(\mathbf{w_i z}))^2) \tag{1.27}$$

(d)

$$\mathbf{w_i(t+1) = w_i(t) + \eta \nabla J} \tag{1.28}$$

where

$$\nabla J = sign(temp) \times temp \tag{1.29}$$

(e) Repeat (b)–(d) until convergence occurs.

ICAdemo.m

```
x1=randn(1,1000);
y1=randn(1,1000);
x2=rand(1,1000);
y2=rand(1,1000);
z1=[0.6 0.3;0.3 0.6]*[x2;y2];
x2=z1(1,:);
y2=z1(2,:);
[E,D]=eig(cov(z1'))
z2=E'*z1;
x3=z2(1,:);
y3=z2(2,:);
x1=(x1-mean(x1))/sqrt(var(x1));
y1=(y1-mean(y1))/sqrt(var(y1));
x2=(x2-mean(x2))/sqrt(var(x2));
y2=(y2-mean(y2))/sqrt(var(y2));
x3=(x3-mean(x3))/sqrt(var(x3));
y3=(y3-mean(y3))/sqrt(var(y3));
figure
subplot(2,2,1)
plot(x1,y1,'*');
subplot(2,2,2)
plot(x2,y2,'*');
subplot(2,2,3)
plot(x3,y3,'*');
%computation of kurtosis of the data
    x1,x2,x3,x4,y2 and y3.
```

```
temp=4e[(Wi^TZ)^3]-12E[Wi^TZ)^2]E[Wi^TZ].
E=[kurt(x1) kurt(y1) kurt(x2) kurt(y2) ...
    kurt(x3) kurt(y3)];
%Absolute value of the kurtosis is minimized as
follows.
%Differentiation of kurtosis of the data with mean=0
%and var=1
W=rand(2,2);
W(:,1)=W(:,1)/sqrt(sum(W(:,1).^2));
W(:,2)=W(:,2)-(W(:,2)'*W(:,1))*W(:,1);
W(:,2)=W(:,2)/sqrt(sum(W(:,2).^2));
x4=x3;
y4=y3;
COL1=[];
COL2=[];
WTRACT=[];
for iter=1:100
X=W'*[x3;y3];
x4=X(1,:);
y4=X(2,:);
W(:,1)=W(:,1)+gradient(x4);
W(:,2)=W(:,2)+gradient(y4);
W(:,1)=W(:,1)/sqrt(sum(W(:,1).^2));
W(:,2)=W(:,2)-(W(:,2)'*W(:,1))*W(:,1);
W(:,2)=W(:,2)/sqrt(sum(W(:,2).^2));
COL1=[COL1 kurt(x4)];
COL2=[COL2 kurt(y4)];
end
subplot(2,2,4)
plot(x4,y4,'*')
figure
subplot(2,1,1)
plot(COL1)
subplot(2,1,2)
plot(COL2)
title('Convergence graph for ICA')
kurt.m
function [res]=kurt(x)
res=(sum(x.^4)/length(x))-3*(sum(x.^2)/length(x))^2;
res=abs(res);
gradient.m
function [res]=gradient(x)
res1=(4*(sum(x.^3)/length(x))-(sum(x.^5)/length(x))...
      + 6*(sum(x.^2)/length(x))*(sum(x.^3)/length(x)));
res=res1*sign(res1);
```

1.6 Maximizing Gaussianity of the Individual Classes for K-LDA

K-LDA helps in mapping the vectors to the higher dimensional space and brings back to the lower dimensional space, such that the vectors within the class are made to come closer to each other and the centroids of various classes are made apart. Construction of the discriminant classifier like Nearest Mean (NM), Nearest Neighbor (NN), and Support Vector Machine (SVM) are made easier if the distribution of the individual classes is Gaussian distributed. This is achieved by maximizing the Gaussianity of the individual classes. The random variable X with zero mean is Gaussian distributed if the absolute of the kurtosis of the random variable is zero. Thus the data after subjected to K-LDA is further linearly transformed using the transformation matrix such that the kurtosis of the data of the individual classes is more Gaussian (by minimizing the absolute value of the kurtosis). It is noted in the case of ICA, and we obtained the transformation matrix such that the data (after transformation) is more non-Gaussian and tried maximizing the absolute value of the kurtosis. Figures 1.12 and 1.13 illustrate the convergence graph for maximizing the Gaussianity of the data obtained after applying KLDA. Figures 1.14, 1.15, and 1.16 illustrate the original data, the data obtained after applying KLDA and the data obtained after maximizing the Gaussianity of the individual classes.

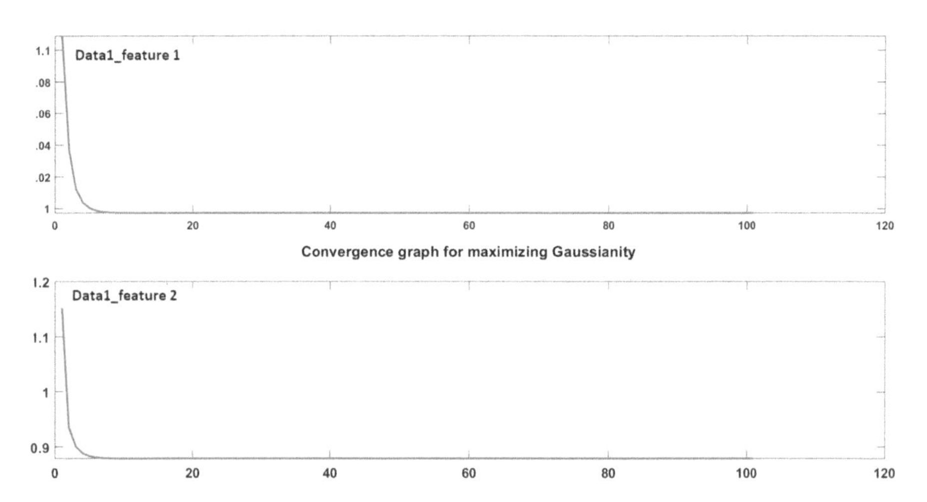

Fig. 1.12 Illustration of the minimization of absolute values of kurtosis for data 1 to obtain ICA basis

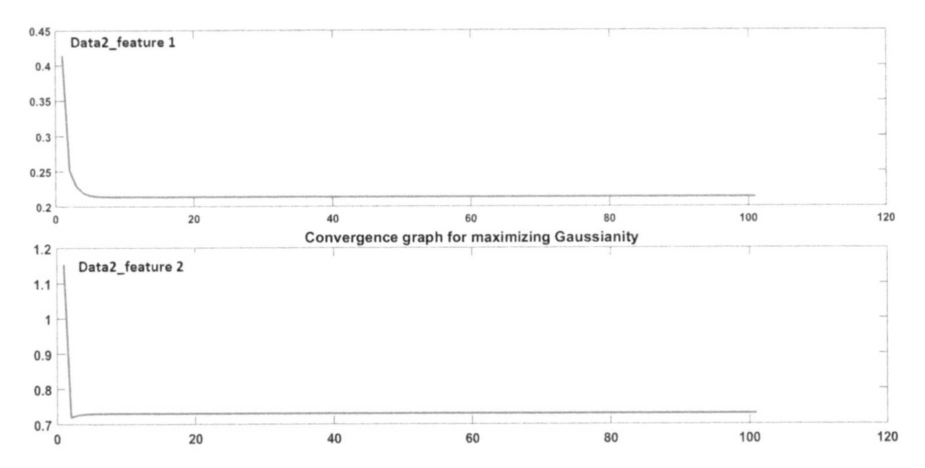

Fig. 1.13 Illustration of the minimization of absolute values of kurtosis for data 2 to obtain ICA basis

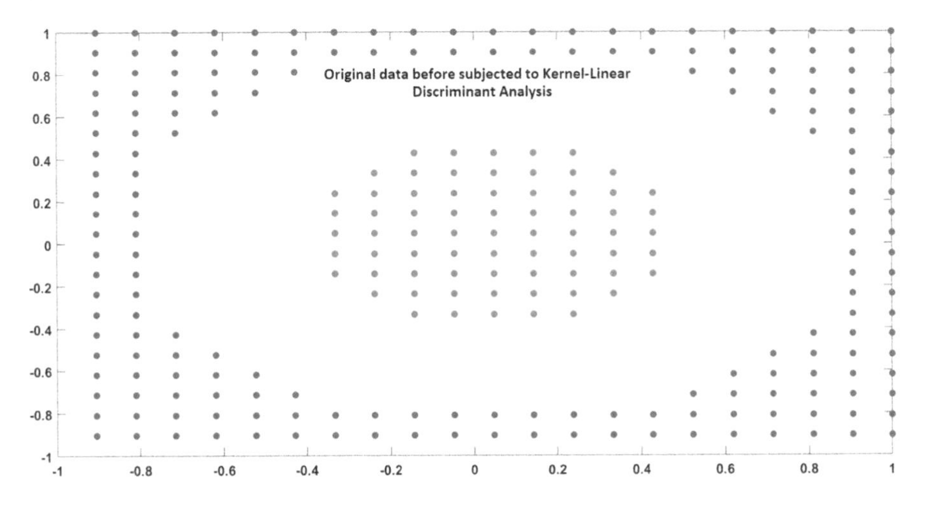

Fig. 1.14 Original data before applying K-LDA

gaussianityafterKLDA.m

```
%Maximizing Gaussianity after applying K-LDA
x=-1:0.1:1;
y=-1:0.1:1;
for i=1:1:length(x)
    for j=1:1:length(y)
z(i,j)=x(i)^2+y(j)^2;
    end
end
[c,h]=contour(x,y,z);
```

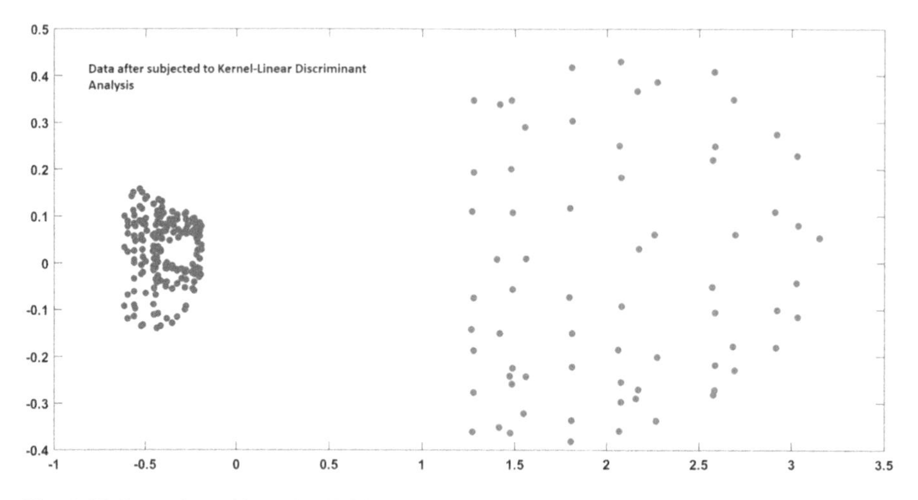

Fig. 1.15 Data after subjected to K-LDA

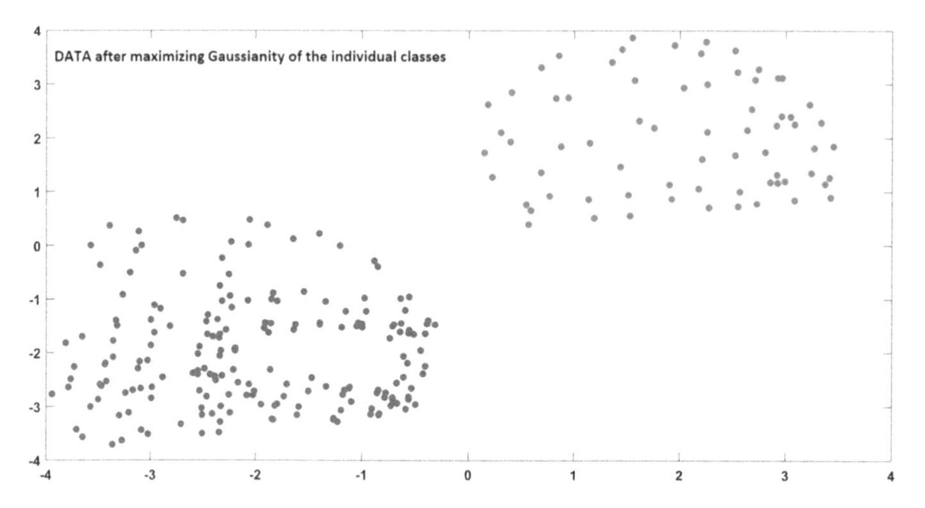

Fig. 1.16 Data after maximizing Gaussianity of the individual classes

```
[x1,y1]=find(0.2<z<0.6);
[x2,y2]=find(z>0.8);
%kernel data
data1=(([x1 y1]/21)*2-1)';
data2=(([x2 y2]/21)*2-1)';
figure
plot(data1(1,:),data1(2,:),'r*')
hold on
plot(data2(1,:),data2(2,:),'b*')
```

```
X=[data1 data2];
sigma=0.2;
split=[1 length(data1) length(data2)]
for i=1:1:size(X,2)
    for j=1:1:size(X,2)
        temp(i,j)=gaussiankernel(X(:,i),X(:,j),sigma);
    end
end
X=temp;
SB=0;
MEANDATA=[];
for i=1:1:length(split)-1
DATA=X(:,split(i):1:split(i)+split(i+1));
n=size(DATA,2);
M=mean(DATA');
temp1=(DATA-repmat(M',1,size(DATA,2)));
temp2=temp1*temp1';
SB=SB+n*temp2;
MEANDATA=[MEANDATA;M];
end
size(MEANDATA)
M=mean(MEANDATA);
temp3=(MEANDATA'-repmat(M',1,size(MEANDATA',2)));
SW=temp3*temp3';
SW=real(SW);
[E,V]=eig(pinv(SW)*SB);
V=diag(real(V))/max(diag(real(V)));
Esel=E(:,1:1:2);
L=size(Esel,2);
Y=real(Esel)'*X;
data1=maxgauss(real(Y(:,1:1:split(2))),[2 2]);
data2=maxgauss(real(Y(:,split(2)+1:1:split(2)+split
(3))),[-2 -2]);
figure
plot(data1(1,:),data1(2,:),'r*')
hold on
plot(data2(1,:),data2(2,:),'b*')
figure
plot(Y(1,1:1:split(2)),Y(2,1:1:split(2)),'r*');
hold on
plot(Y(1,split(2)+1:1:split(2)+split(3)),Y(2,split(2)
+1:1:split(2)+split(3)),'b*');
```

maxgauss.m

```
%maxgauss.m
function [res]=maxgauss(x,M)
CX=cov(x');
[E,D]=eig(CX);
y=real(E)*x;
y=y-repmat(mean(y')',1,length(y));
x1=y(1,:)/sqrt(var(y(1,:)));
y1=y(2,:)/sqrt(var(y(2,:)));
%Absolute value of the kurtosis is minimized as
follows.
%Differentiation of kurtosis of the data with mean=0
%and var=1
W=rand(2,2);
W(:,1)=W(:,1)/sqrt(sum(W(:,1).^2));
W(:,2)=W(:,2)-(W(:,2)'*W(:,1))*W(:,1);
W(:,2)=W(:,2)/sqrt(sum(W(:,2).^2));
COL1=[kurt(x1)];
COL2=[kurt(y1)];
WTRACT=[];
x2=x1;y2=y1;
for iter=1:100
W(:,1)=W(:,1)-gradient(x2);
W(:,2)=W(:,2)-gradient(y2);
W(:,1)=W(:,1)/sqrt(sum(W(:,1).^2));
W(:,2)=W(:,2)-(W(:,2)'*W(:,1))*W(:,1);
W(:,2)=W(:,2)/sqrt(sum(W(:,2).^2));
X=W'*[x1;y1];
x2=X(1,:);
y2=X(2,:);
COL1=[COL1 kurt(x2)];
COL2=[COL2 kurt(y2)];
end
figure
subplot(2,1,1)
plot(COL1)
subplot(2,1,2)
plot(COL2)
title('Convergence graph for maximizing Gaussianity')
CZ=cov(real([x2;y2])')
[E,D]=eig(real(CZ));
res=real(E)*(real(D)^(1/2))*[x2;y2];
res=res+repmat(M',1,length(res));
figure
```

```
subplot(2,2,1)
plot(x(1,:),x(2,:),'*')
subplot(2,2,2)
plot(x1,y1,'*')
subplot(2,2,3)
plot(x2,y2,'*')
subplot(2,2,4)
plot(real(res(1,:)),real(res(2,:)),'*')
```

Chapter 2
Linear Classifier Techniques

2.1 Discriminative Supervised Classifier

Discriminative supervised classifier involves identifying the hyperplane such that $\mathbf{w}_i^T\mathbf{x} + b > 0$, if $\mathbf{x}$ belongs to the ith class. For instance, consider three classes red (asterisk), blue (diamond), and black (circle) vectors in 2D space as shown in Fig. 2.1. In this case, the region $y + 5x - 3 \geq 0$ belongs to class 1, $y - 2x + 3 \geq 0$ belongs to class 2, and $-y - 3x + 4 \geq 0$ belongs to class 3. Each region is represented as "+" in the figure. It is seen in the figure that more than one + are available in some partition regions. If the arbitrary vector (under test) lies in the region, indicated by single + symbol, we can classify that vector belongs to one among the three classes. But if the arbitrary vector (under test) lies in the region, (plus) indicated by more than one + (plus), we cannot declare those vector into one among the three classes. These regions are known as in-determinant regions (IR). IR is represented with white regions in Fig. 2.1. It is also observed that the middle region is indicated with three plus (+) symbols, representing that if the vector is in this region, we cannot declare it into one among the three classes. This is known as Type 1 discrimination. We obtain the collinear lines (three) using the constructed three lines as $y - 9x + 9 = 0$, $y + 13x - 18 = 0$, and $-y - 2x + 4.5 = 0$ (refer Fig. 2.2). In this case all the three lines meet at a point and we still have IR. But the middle IR is not present. This is known as Type 2 discrimination.

discriminationdemo.m

```
%Type 1
xp=-4:0.1:4
y1p=-5*xp+3;
y2p=2*xp-3;
y3p=-3*xp+4;
DATA=[];
```

© Springer Nature Switzerland AG 2020

E. S. Gopi, *Pattern Recognition and Computational Intelligence Techniques Using Matlab*, Transactions on Computational Science and Computational Intelligence,
https://doi.org/10.1007/978-3-030-22273-4_2

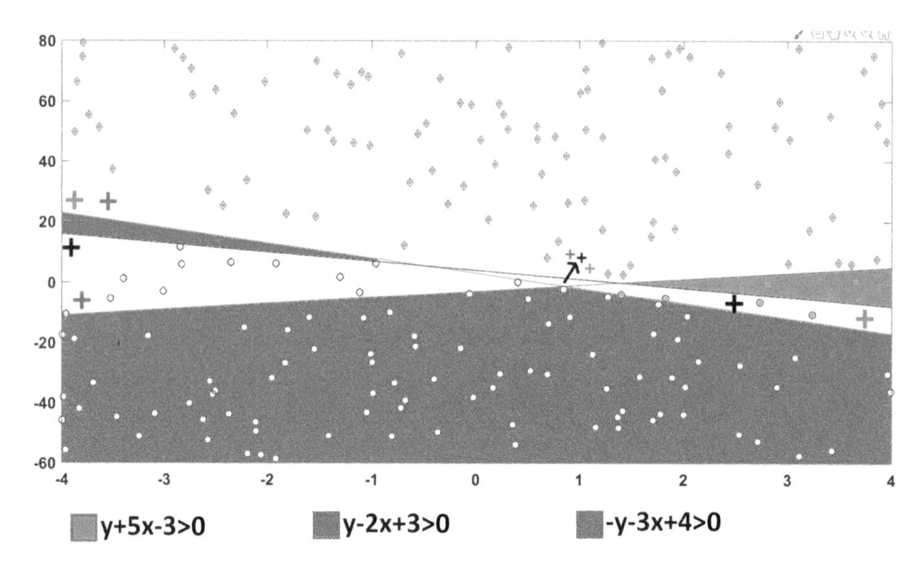

Fig. 2.1 Illustration of Type 1 discrimination (class i versus others)

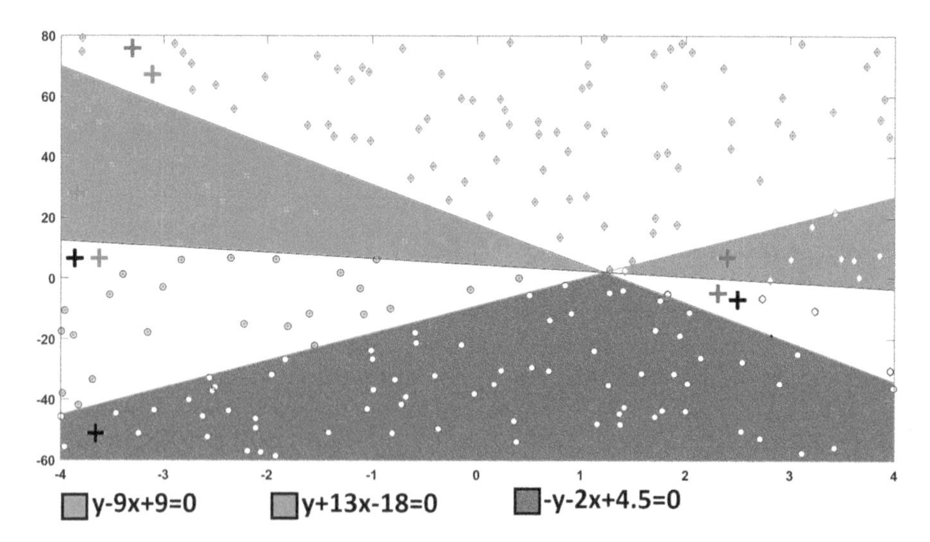

Fig. 2.2 Illustration of Type 2 discrimination (class i versus others)

```
for i=1:1:200
    x=rand*8-4;
    y=rand*140-60;
DATA=[DATA;x y];
end
figure(1)
plot(xp,y1p,'r-')
```

```
hold on
plot(xp,y2p,'b-')
plot(xp,y3p,'k-')
plotdata(DATA,5,1,-3,1,'r*')
plotdata(DATA,-2,1,3,1,'bd')
plotdata(DATA,-3,-1,4,1,'ko')
%%%%%%%%%%%%%%%%%%%%%%%%%%%%%%%%%%%%
xp=-4:0.1:4
z1p=9*xp-9;
z2p=-13*xp+18;
z3p=-2*xp+4.5;
figure(2)
plot(xp,z1p,'r-')
hold on
plot(xp,z2p,'b-')
plot(xp,z3p,'k-')
plotdata(DATA,-9,1,9,2,'r*')
plotdata(DATA,13,1,-18,2,'bd')
plotdata(DATA,-2,-1,4.5,2,'ko')
```

plotdata.m

```
function [res]=plotdata(data,a,b,c,n,s)
for i=1:1:length(data)
    x=data(i,1);
    y=data(i,2);
    if((a*x+b*y+c)>0)
        figure(n)
        hold on
        plot(x,y,s)
    end
end
res=1;
```

2.2 Nearest Mean (NM) Classifier

In this technique, the mean of the individual classes is identified. To declare the arbitrary vector into one among the finite number of classes, the Euclidean distance between the arbitrary vector and the mean of the individual classes is computed. The arbitrary vector is declared as the one belonging to the particular class that has the minimum Euclidean distance.

2.2.1 Mean Distance Versus Mahalanobis Distance

We model the individual class conditional probability density function (ccpdf) as the multivariate Gaussian density function with the specified mean vector $\mathbf{m_i}$ and the co-variance matrix $\mathbf{C}_i$. Assign the arbitrary vector $\mathbf{v}$ belongs to class k if $k = arg_i min(\mathbf{v} - \mathbf{m_i})^T \mathbf{C}_i^{-1}(\mathbf{v} - \mathbf{m_i})$. $(\mathbf{v} - \mathbf{m_i})^{\mathbf{T}} \mathbf{C}_i^{-1}(\mathbf{v} - \mathbf{m_i})$ is known as Mahalanobis distance. This is inversely proportional to the ccpdf of the particular class. It is understood that we assign the arbitrary vector to the class that has the maximum class conditional probability density function. This is illustrated in Figs. 2.3, 2.4, 2.5 and 2.6.

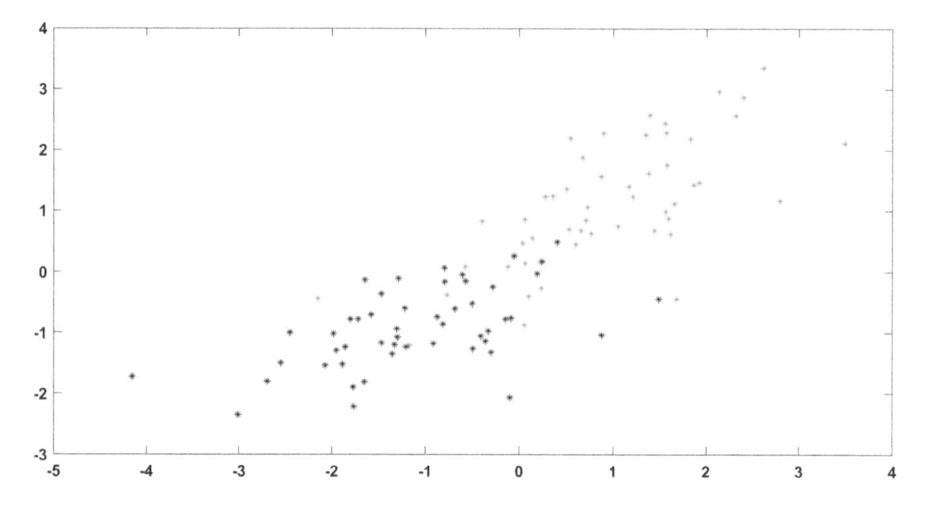

Fig. 2.3 Training set used to compute Euclidean distance and Mahalanobis distance

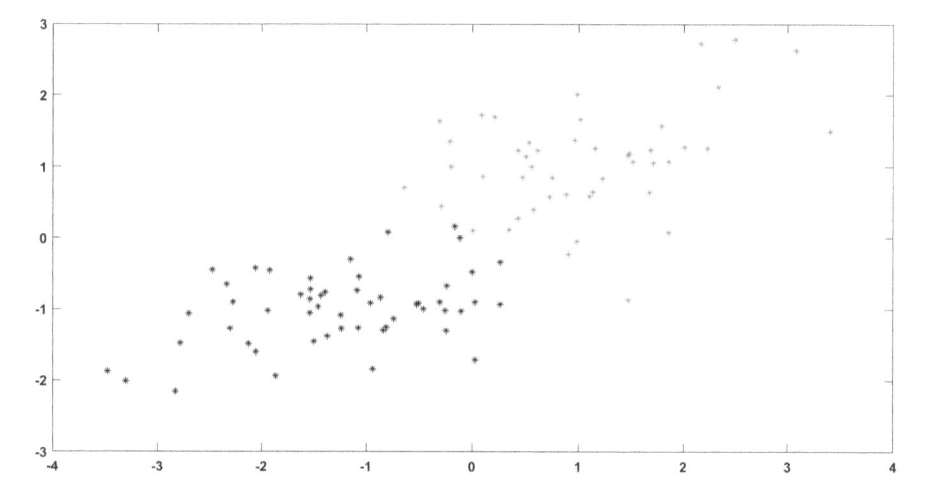

Fig. 2.4 Class indeed obtained using Euclidean distance with mean

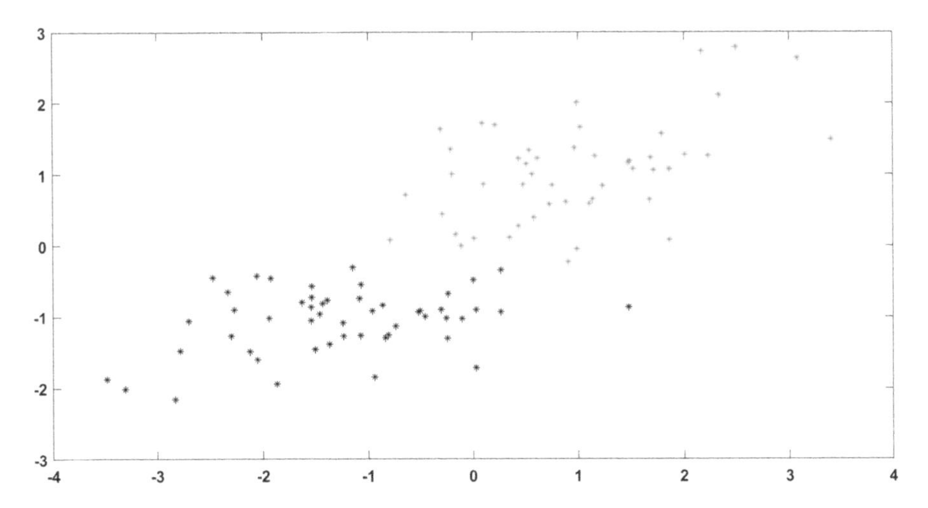

Fig. 2.5 Class index obtained using Mahalanobis distance for the test data

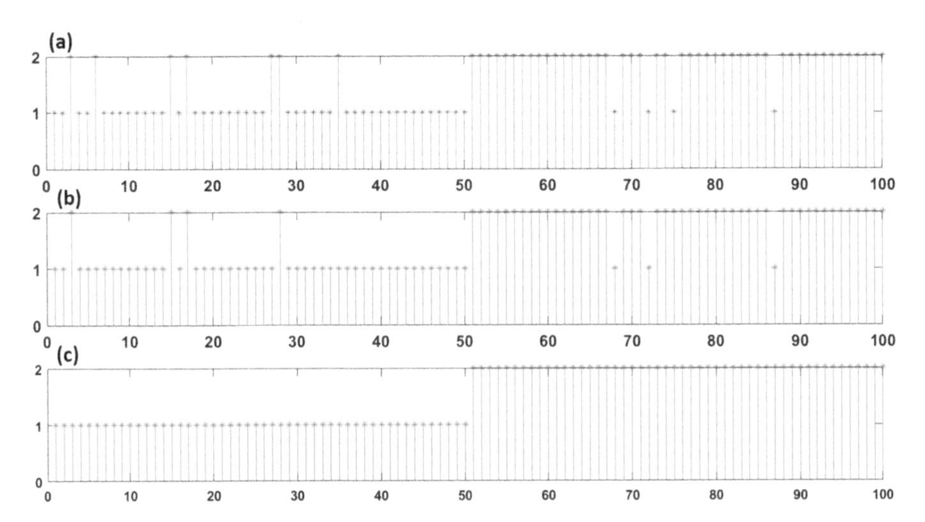

Fig. 2.6 (**a**) Class label obtained using Euclidean distance. (**b**) Class label obtained using Mahalanobis distance. (**c**) Actual class label

nearestmean.m

```
%Nearest mean
data=randn(2,100);
M1=[0.9 0.3;0.4 0.9];
M2=[0.9 0.2;0.2 0.6];
x=M1*data+repmat([1 1]',1,100);
y=M2*data+repmat([-1 -1]',1,100);
TRAINDATA=[x(:,1:1:50) y(:,1:1:50)];
```

```
C1=cov(x(:,1:1:50)');
TESTDATA=[x(:,51:1:100) y(:,51:1:100)];
C2=cov(y(:,1:1:50)');
figure
plot(x(1,1:1:50),x(2,1:1:50),'r*')
hold on
plot(y(1,1:1:50),y(2,1:1:50),'k*')
%Euclidean distance
MEAN1=mean(x(:,1:1:50)');
MEAN2=mean(y(:,1:1:50)');
INDEX=[];
S1=sum((TESTDATA-repmat(MEAN1',1,100)).^2);
S2=sum((TESTDATA-repmat(MEAN2',1,100)).^2);
INDEX1=[];
for i=1:1:100
if(S1(i)<S2(i))
    INDEX1=[INDEX1 1];
else
    INDEX1=[INDEX1 2];
end
end
%Mahanobolis distance
INDEX2=[];
S1=sum((TESTDATA-repmat(MEAN1',1,100))'*inv(C1)*
(TESTDATA-repmat(MEAN1',1,100)));
S2=sum((TESTDATA-repmat(MEAN2',1,100))'*inv(C2)*
(TESTDATA-repmat(MEAN2',1,100)));
INDEX2=[];
for i=1:1:100
if(S1(i)<S2(i))
    INDEX2=[INDEX2 1];
else
    INDEX2=[INDEX2 2];
end
end
POS1=length(find((INDEX1-[ones(1,50) ones(1,50)*2])
==0));
POS2=length(find((INDEX2-[ones(1,50) ones(1,50)*2])
==0));
figure
subplot(3,1,1)
stem(INDEX1,'b*')
subplot(3,1,2)
stem(INDEX2,'r*')
subplot(3,1,3)
```

```
stem([ones(1,50) ones(1,50)*2],'r*')
figure
for i=1:1:100
    if(INDEX1(i)==1)
plot(TESTDATA(1,i),TESTDATA(2,i),'r*')
hold on
    else
plot(TESTDATA(1,i),TESTDATA(2,i),'k*')
   end
end
figure
for i=1:1:100
    if(INDEX2(i)==1)
plot(TESTDATA(1,i),TESTDATA(2,i),'r*')
hold on
    else
plot(TESTDATA(1,i),TESTDATA(2,i),'k*')
    end
end
```

2.3 Nearest Neighbor (NN) Classifier

In this technique, we look for class index of the k vectors nearest to the vector under test. We assign the arbitrary vector to the class with the index with largest count. This is illustrated in Figs. 2.7 and 2.8.

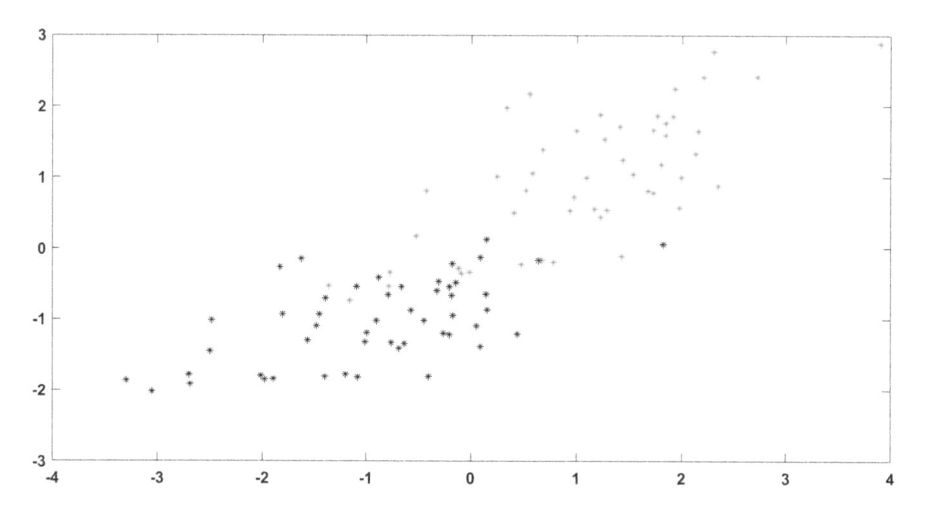

Fig. 2.7 Training data used in NN demonstration

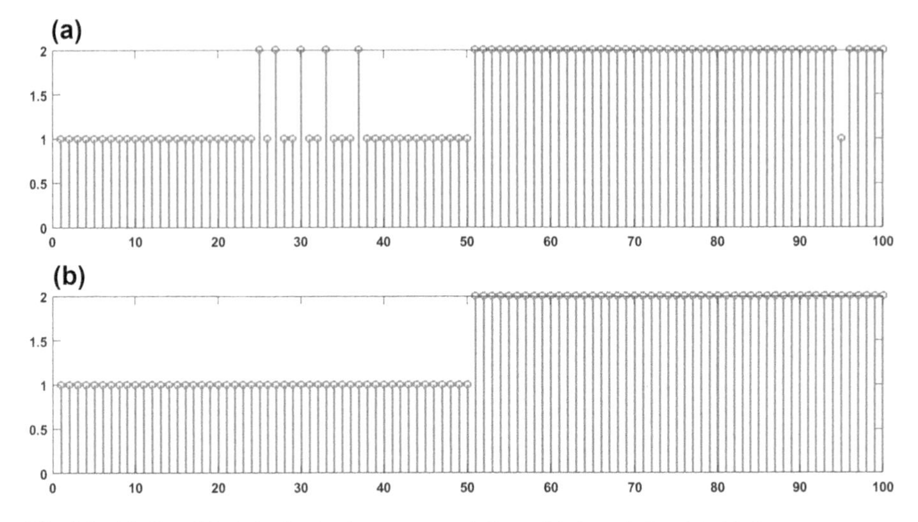

Fig. 2.8 (**a**) Class label obtained using nearest neighbor. (**b**) Actual class label index

nearestneighbour.m

```
Neighbour
data=randn(2,100);
M1=[0.9 0.3;0.4 0.9];
M2=[0.9 0.2;0.2 0.6];
x=M1*data+repmat([1 1]',1,100);
y=M2*data+repmat([-1 -1]',1,100);
TRAINDATA=[x(:,1:1:50) y(:,1:1:50)];
TESTDATA=[x(:,51:1:100) y(:,51:1:100)];
figure
plot(x(1,1:1:50),x(2,1:1:50),'r*')
hold on
plot(y(1,1:1:50),y(2,1:1:50),'k*')
INDEX=[];
for i=1:1:100
    S=sum((TRAINDATA-TESTDATA(:,i)).^2);
    [P,Q]=sort(-1*S);
    COL=Q(1:1:10);
    if((length(find(COL<50)))<(length(find(COL>50))))
        INDEX=[INDEX 1];
    else
        INDEX=[INDEX 2];
    end
end
figure
subplot(2,1,1)
stem(INDEX)
```

```
subplot(2,1,2)
stem([ones(1,50)  ones(1,50)*2] )
```

2.4 Perceptron Model

The perceptron algorithm is used to identify the line/plane/hyperplane of the form $\mathbf{w}^T\mathbf{x}$ that divides the two class. The vector under test $\mathbf{v}$ is said to be the one belongs to class 1 if $\mathbf{w}^T[\mathbf{v}; 1] > 0$ and class 2 if $\mathbf{w}^T[\mathbf{v}; 1] < 0$. Let us consider the training vectors are in 2D space. Given the training set $\{x_1, x_2, ..., x_N\}$ with known indices, the objective function that is used to optimize $\mathbf{w}$ is represented in (2.1), where we assign $t_k = 1$ if the vector $\mathbf{x}_k \in$ class 1 and -1 if $\mathbf{x}_k \in$ class 2.

$$J = -\sum_{i=1}^{i=N, i\in M} \mathbf{w}^T[\mathbf{x}_i; 1]t_k \tag{2.1}$$

is minimized. M is the misclassification set. The step involved in updating the weight vector (in $(t+1)^{\text{th}}$ iteration)) is given as follows:

$$\mathbf{w}(t+1) = \mathbf{w}(t) + \eta[\mathbf{x_k}; 1]t_k \tag{2.2}$$

Equation (2.2) is realized for the individual misclassification vector one after another, where $x_k \in M$.

The $J(t+1)$ and $J(t)$ are computed using $w(t+1)$ and $w(t)$, respectively, and are given as follows:

$$J(t+1) = -\sum_{i=1}^{i=N, i\in M} \mathbf{w}(t+1)^T[\mathbf{x}_i; 1]t_k \tag{2.3}$$

$$= -\sum_{i=1}^{i=N, i\in M} (w(t) + \eta[\mathbf{x}_k; 1]t_k)^T[\mathbf{x}_i; 1]t_k \tag{2.4}$$

$$\Rightarrow -\sum_{i=1}^{i=N, i\in M} (\mathbf{w}(t)^T[\mathbf{x}_i; 1]t_k - \eta \sum_{i=1}^{i=N, i\in M} [\mathbf{x}_k; 1]^T[\mathbf{x}_k; 1](t_k)^2 \tag{2.5}$$

$$\Rightarrow J(t) - (+ve\ quantity) \tag{2.6}$$

Hence $J(t+1)$ is lesser than $J(t)$, and minimization is achieved after every iteration. This is illustrated in Figs. 2.9, 2.10, and 2.11.

perceptron.m

```
%Perceptron model
%Given the 2D vectors, identifying the line
%that divides the vectors into two groups using
```

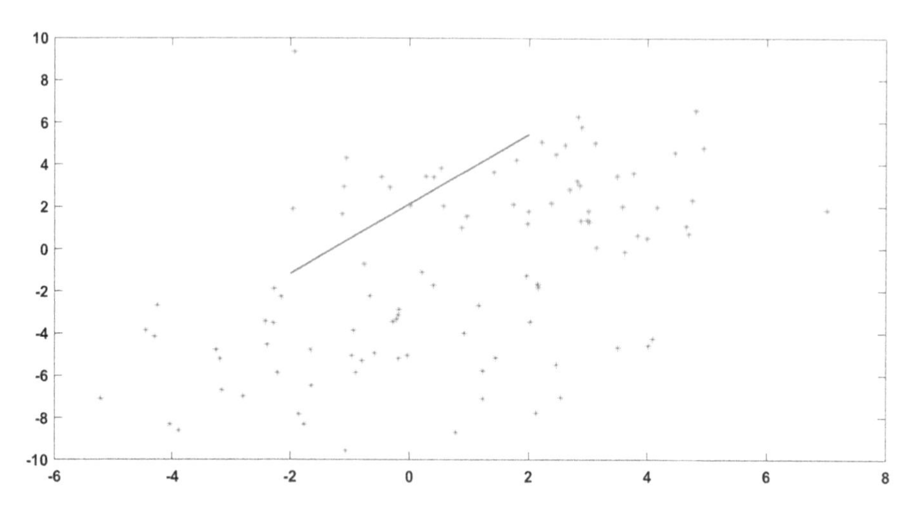

Fig. 2.9 Initial guess of the partition line

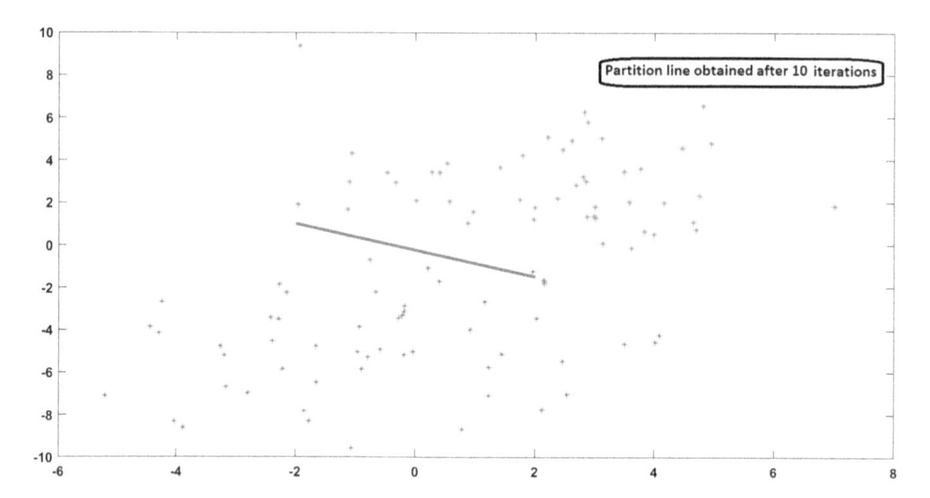

Fig. 2.10 Partition line obtained after 10 iterations using perceptron algorithm

```
perceptron
%is as given below.
DATA1=randn(2,50)*2+repmat([2 3]',1,50);
DATA2=randn(2,50)*2+repmat([-1 -5]',1,50);
DATA=[DATA1 DATA2];
t=[ones(1,50) ones(1,50)*(-1)];
figure
plot(DATA1(1,:),DATA1(2,:),'r*')
hold on
plot(DATA2(1,:),DATA2(2,:),'b*')
```

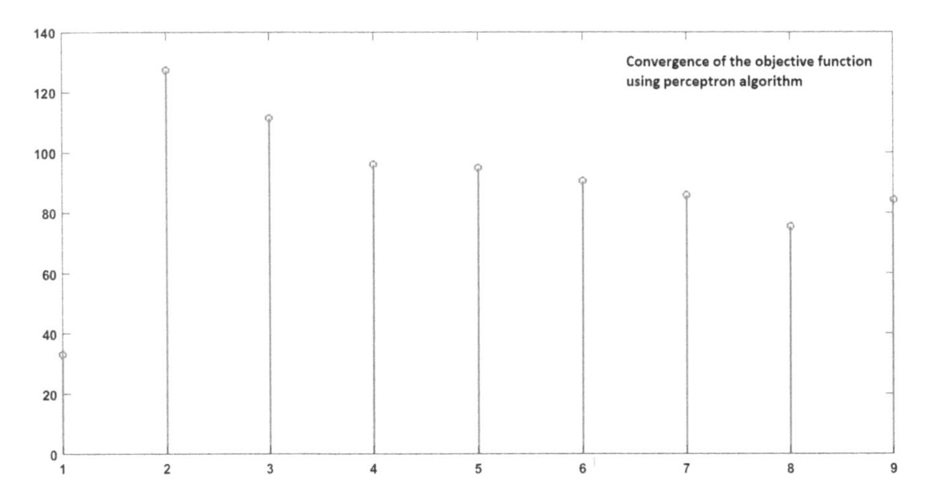

Fig. 2.11 Convergence obtained using perceptron algorithm

```
%Initialize the weight
w=rand(1,3)*2-1;
figure(1)
x=-2:0.1:2
y=(ones(1,length(x))*(-w(3))-w(1)*x)/w(2);
hold on
plot(x,y,'-')
J=[];
for iteration=1:1:10
INDEX=DATA'*w(1:2)'+w(3);
%Find the vector with error. We actually assign t
values as 1
%for 1 to 50 vectors and -1 for 51 to 100 vectors
%The objective function is computed as follows.
%Identify the error vector
S=sign(INDEX);
[P1,Q1]=find(S(1:1:50)==-1);
[P2,Q2]=find(S(51:1:100)==1);
E=[P1;P2+50];
%Update weights
for i=1:1:length(E)
    w=w+rand*[DATA(:,E(i));1]'*t(E(i));
end
s=0;
for i=1:1:length(E)
    s=s-w*[DATA(:,E(i));1]*t(E(i));
end
```

```
J=[J s];
end
figure(2)
plot(DATA1(1,:),DATA1(2,:),'r*')
hold on
plot(DATA2(1,:),DATA2(2,:),'b*')
x=-2:0.1:2
y=(ones(1,length(x))*(-w(3))-w(1)*x)/w(2);
hold on
plot(x,y,'-')
figure(3)
stem(J(2:1:10))
```

2.5 Support Vector Machine (SVM)

Support Vector Machine (SVM) involves identifying the partition line in 2D space that separates the vectors (partition line) belonging to the class 1 and the vectors belonging to the class 2. The index of the vectors belonging to the particular classes is known. Let t_k be the value assigned to the individual vectors. It takes either $+1$, if the vector belongs to class 1, or -1, if the vector belongs to class 2. The partition line is described as $w_1x + w_2y + b = 0$ in the x–y plane. Consider another two lines parallel to the partition line. The equation of those lines is as given below. The requirement is to obtain the optimal values for w_1 and w_2 and b such that the perpendicular distance between the lines is maximized as follows:

$$w_1x + w_2y + b - 1 = 0 \tag{2.7}$$

$$w_1x + w_2y + b + 1 = 0 \tag{2.8}$$

The projection of the arbitrary point (p, q) to the lines $w_1x + w_2y + b - 1 = 0$ and $w_1x + w_2y + b + 1 = 0$ is given as follows:

$$\frac{w_1p + w_2q + b - 1}{\sqrt{w_1^2 + w_2^2}} \tag{2.9}$$

$$\frac{w_1p + w_2q + b + 1}{\sqrt{w_1^2 + w_2^2}} \tag{2.10}$$

The perpendicular distance between these lines is computed as follows:

$$\frac{2}{\sqrt{w_1^2 + w_2^2}} \tag{2.11}$$

Thus to maximize the perpendicular distance between the lines, $||\mathbf{w}||$ needs to be minimized. Given the arbitrary vector $\mathbf{x}_k = [x_{1k}\ x_{2k}]$ belonging to class 1 satisfy the condition $w_1x + w_2y + b < -1$ and if the vector belongs to class 2, it satisfies the condition $w_1x + w_2y + b > 1$. We assign the value for the variable t_k for the kth training vector as -1 or $+1$ if it belongs to for class 1 and class 2 respectively. Thus the arbitrary vector $\mathbf{x}_k$ satisfies the following condition.

$$(w_1x_{1k} + w_2x_{2k} + b)t_k > 1 \tag{2.12}$$

The objective function for SVM is structured as follows:

$$J = \frac{w_1^2 + w_2^2}{2} \tag{2.13}$$

with the constraints that

$$\sum_{k=1}^{k=N}((w_1x + w_2y + b)t_k) \geq 1 \tag{2.14}$$

We use the Lagrangian technique to solve this problem. The Lagrangian equation is structured as follows:

$$J = \frac{w_1^2 + w_2^2}{2} - \sum_{k=1}^{k=N}\lambda_k((w_1x_{1k} + w_2x_{2k} + b)t_k - 1) \tag{2.15}$$

with the following Kuhn–Tucker conditions:

1. $\lambda_k > 0$
2. $\lambda_k((w_1x_{1k} + w_2x_{2k} + b)t_k - 1) = 0$

2.6 Formulating the Dual Problem

Differentiating (2.10) with respect to w_1 and w_2, we get the following:

$$w_1 = \sum_{k=1}^{k=N}\lambda_k x_{1k} t_k \tag{2.16}$$

$$w_2 = \sum_{k=1}^{k=N}\lambda_k x_{2k} t_k \tag{2.17}$$

$$\Rightarrow w = \sum_{k=1}^{k=N}\lambda \mathbf{x}_k t_k \tag{2.18}$$

Substituting (2.16) and (2.17) in (2.15), we get the dual problem as follows:

$$\frac{\mathbf{w}^T\mathbf{w}}{2} - \sum_{k=1}^{k=N} \lambda_k((\mathbf{w}^T\mathbf{x_k} + b)t_k - 1) \tag{2.19}$$

The first term is given as follows:

$$\frac{\left(\sum_{m=1}^{m=N} \lambda_m \mathbf{x_m} t_m\right)^T \left(\sum_{n=1}^{n=N} \lambda_n \mathbf{x_n} t_n\right)}{2} \tag{2.20}$$

The second term is given as follows:

$$\sum_{k=1}^{k=N} \lambda_k \left(\left(\left(\sum_{m=1}^{m=N} \lambda_m \mathbf{x_m} t_m\right)^T \mathbf{x_k} + b\right) t_k - 1\right) \tag{2.21}$$

The above structured optimization function is the maximization problem without constraints. We need to get the solution λ's and b that maximizes the formulated objective function with the Kuhn–Tucker conditions.

Differentiating with respect to b and equating to 0, we get the following:

$$\sum_{k=1}^{k=N} \lambda_k t_k = 0 \tag{2.22}$$

Differentiating (2.21) with respect to λ_r, equating to 0, and solving for w (with KKT condition), we get the following:

$$\mathbf{M}\boldsymbol{\lambda} = 1_N \tag{2.23}$$

where M is the Gram-matrix with i^{th} column vectors are obtained as

$$[\mathbf{x_1}^T\mathbf{x_i}\ \mathbf{x_2}^T\mathbf{x_i}\ \cdots \mathbf{x_N}^T\mathbf{x_i}] \tag{2.24}$$

Combining both the equations, we get the following:

$$[\mathbf{M};\ T][\boldsymbol{\lambda}] = [1_N; 0] \tag{2.25}$$

where $T = [t_1\, t_2 \cdots t_N]^T$. The set of $\mathbf{x_k}$'s corresponding to the kth column of the matrix M with $\lambda_k > 0$ is called support vectors. These vectors satisfy the following condition:

$$(w_1 x_{1k} + w_2 x_{2k} + b)t_k = 1 \tag{2.26}$$

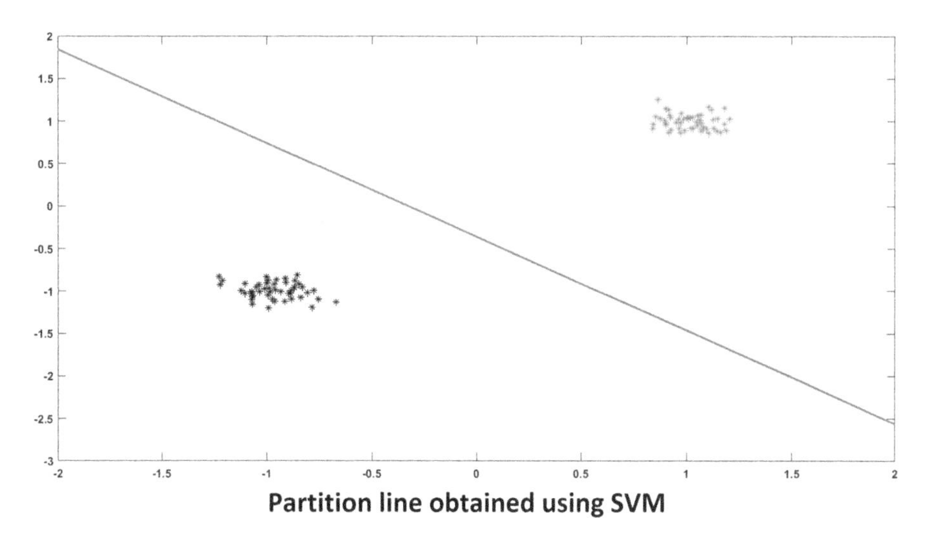

Fig. 2.12 Demonstration of obtaining the partition line using SVM

Multiplying t_k on both sides, and solving for b, we get the following: $b = t_k - w_1 x_{1k} - w_2 x_{2k}$. Thus b is estimated as the averaging over all support vectors, i.e.,

$$b = \frac{1}{r} \sum_{k=1}^{k=N} (t_k - w_1 x_{1k} - w_2 x_{2k}) \tag{2.27}$$

where K $\acute{\epsilon}$ support vectors and r is the number of support vectors. The typical partition line described by $\mathbf{w}$ and b for the given training set is illustrated in Fig. 2.12.

2.6.1 *Kernel Trick for SVM*

If the classes are well separated, finding out the partition line that divides the classes is obtained using the steps as described in the previous section. If the classes are not well separated, we need to map the vectors to the higher dimensional space. It is assumed that when the vectors are mapped to the higher dimensional space, the vectors are well separated and hence there exists the hyperplane that partitions the classes in the higher dimensional space. This is done using "Kernel Trick" without the actual map to the higher dimensional space as follows:

- Let the vector in the lower dimensional space is mapped to the higher dimensional space using the transformation $\phi(\mathbf{x})$ (ϕ consists of m basis functions).

- As the matrix **M** consists of the elements computed using inner product, we replace the ith column with the inner-product vectors computed using the vectors mapped to the higher dimensional space using $\phi(\mathbf{x})$.
- Thus kernel-based SVM consists of constructing the matrix M_ϕ, with i^{th} column vector is represented as follows: $[\phi(\mathbf{x})^T\phi(\mathbf{x1})\ \phi(\mathbf{x})^T\phi(\mathbf{x2})\ \cdots\phi(\mathbf{x})^T\phi(\mathbf{xN})]^T$
- If we define $k(\mathbf{x}, \mathbf{x1}) = \phi(\mathbf{x})^T\phi(\mathbf{x1})$, where k is the kernel function. By choosing the proper kernel function, we are performing the SVM in the higher dimensional space without actual mapping to the higher dimensional space using ϕ. This is known as "Kernel Trick."
- Compute the λ's using the following equation: $[M_\phi; T]\lambda = [1_N[0]$.
- The set of vectors $\phi(\mathrm{x_k})'\Delta$ (corresponding to k^{th} column of the matrix M_ϕ), with $\lambda_\mathrm{k} > 0$ are called support vectors.
- The transformation matrix in the higher dimensional space is computed as follows:

$$\mathbf{w}_\phi = \sum_{k=1,k\in S}^{k=N} \lambda_k\phi(\mathbf{x}_k)t_k \tag{2.28}$$

where S is the support vector set, i.e., $\phi(\mathrm{x_i})$ corresponding to $\lambda_i > 0$.
- The bias vector b_ϕ satisfies the equation $(\mathbf{w}_\phi^T\phi(x_l) + b_\phi)t_l = 1$ for every support vectors. By multiplying t_1 on both sides of the equation, b_ϕ is obtained as follows:

$$\mathbf{w}_\phi^T\phi(\mathbf{x}_l) + b_\phi = t_1 \tag{2.29}$$

$$\Rightarrow b_\phi = t_1 - \mathbf{w}_\phi^T\phi(\mathbf{x}_l) \tag{2.30}$$

$$\Rightarrow b_\phi = t_l - \left(\sum_{k=1,k\in S}^{k=N} \lambda_k\phi(\mathbf{x}_k)t_k\right)^T \phi(\mathbf{x}_l) \tag{2.31}$$

$$\Rightarrow b_\phi = t_l - \sum_{k=1,k\in S}^{k=N} \lambda_k t_k(\phi(\mathbf{x}_k)^T\phi(\mathbf{x}_l) \tag{2.32}$$

$$\Rightarrow b_\phi = t_l - \sum_{k=1,k\in S}^{k=N} \lambda_k t_k k(\mathbf{x}_k, \mathbf{x}_l) \tag{2.33}$$

- The b_ϕ in (2.33) is true for all l, and hence b_ϕ is estimated as the averaging of b_ϕ obtained over all the support vectors as follows:

$$b_\phi = \frac{1}{K}\sum_{l=1,l\in S}^{l=N}(t_l - \sum_{k=1,k\in S}^{k=N} \lambda_k t_k k(\mathbf{x}_k, \mathbf{x}_l)) \tag{2.34}$$

- K is the number of support vectors.
- To check whether the arbitrary vector $PL(v)$ belongs to class 1 or class 2, we use the partition (line/plane/ hyperplane) $PL = \mathbf{w}_{\phi}^{T}\phi(\mathbf{v}) + b_{\phi}$. If $PL(v) > 1$, we declare the vector v belongs to class 1. If $PL(v) < -1$, we declare the vector v belongs to class 2. $PL(v)$ is computed using kernel function as described follows:

$$PL(v) = \left(\sum_{k=1, k \in S}^{k=N} \lambda_k \phi(\mathbf{x}_k) t_k \right)^T \phi(\mathbf{v}) + b_{\phi} \tag{2.35}$$

$$PL(v) = \sum_{k=1, k \in S}^{k=N} \lambda_k t_k k(\mathbf{x}_k, \mathbf{v}) + b_{\phi} \tag{2.36}$$

- The illustration using the training vectors in the 2D space is illustrated as follows:

svmdemo.m

```
Xtrain=randn(2,50)*0.1+1;
Ytrain=randn(2,50)*0.1-1;
DATAtrain=[Xtrain Ytrain];
Xtest=randn(2,50)*0.1+1;
Ytest=randn(2,50)*0.1-1;
DATAtest=[Xtest Ytest];
t=[ones(1,50) ones(1,50)*(-1)];
v=10;
%Formulation of Gram-matrix
for i=1:1:100
    for j=1:1:100
        k(i,j)=gaussiankernel(DATAtrain(:,i),
        DATAtrain(:,j),v)*t(i)*t(j);
    end
end
M=[k;t];
W=pinv(M)*ones(101,1);
%Selection of Lagrangean constant greater
%than zero (Support vectors)
[p,q]=find(W>0);
%b is obtained as follows.
temp=0;
for i=1:1:length(p)
    s1=t(q(i));
    s=0;
    for j=1:1:length(p)
    s=s+k(q(i),q(j))*t(q(j))
    end
    temp=temp+s1-s;
```

```
end
temp=temp/length(q);
%Check for the training set
RES=[];
for i=1:1:100
    s=0;
    for j=1:1:length(q)
s=s+gaussiankernel(DATAtrain(:,i),DATAtrain(:,q(j)),
v)*t(q(j))
    end
    s=s+temp;
RES=[RES sign(s)];
end
%Checking for the testing test
RES=[];
for i=1:1:100
    s=0;
    for j=1:1:length(q)
s=s+gaussiankernel(DATAtest(:,i),DATAtrain(:,q(j)),
v)*t(q(j))
    end
    s=s+temp;
RES=[RES sign(s)];
end
L1=length(find(RES(1:1:50)==1));
L2=length(find(RES(51:1:100)==- 1));
POS1=(L1+L2);
figure(1)
for i=1:1:50
    plot(DATAtest(1,i),DATAtest(2,i),'r*')
    hold on
end
for i=51:1:100
    plot(DATAtest(1,i),DATAtest(2,i),'k*')
    hold on
end
%%%%%%%%%%%%%%%%%%%%%%%%%%%%%%%%%%%%%%%%%%%%%%%%%%%%%
%Formulation of Gram-matrix
for i=1:1:100
    for j=1:1:100
        k(i,j)=innerproduct(DATAtrain(:,i),DATAtrain
        (:,j))*t(i)*t(j);
    end
end
M=[k;t];
```

```
W=pinv(M)*ones(101,1);
%Selection of Lagrangean constant greater
%than zero (Support vectors)
[p,q]=find(W>0);
%b is obtained as follows.
temp=0;
for i=1:1:length(p)
    s1=t(q(i));
    s=0;
    for j=1:1:length(p)
    s=s+k(q(i),q(j))*t(q(j))
    end
    temp=temp+s1-s;
end
temp=temp/length(q);
s=0;
for i=1:1:length(q)
s=s+DATAtrain(:,q(i))*W(q(i))*t(q(i));
end
w=s;
b=0;
for i=1:1:length(p)
    b=b+t(q(i))-w(1)*DATAtrain(1,q(i))+w(2)*DATAtrain
    (2,q(i))
end
b=b/length(q);
x=-2:0.1:2
y=(ones(1,length(x))*(-b)-w(1)*x)/w(2);
hold on
plot(x,y,'-')
%Check for the training set
RES=[];
for i=1:1:100
    s=w(1)*DATAtrain(1,i)+w(2)*DATAtrain(2,i)+b;
    RES=[RES sign(s)];
end
%Checking for the testing test
RES=[];
for i=1:1:100
s=w(1)*DATAtest(1,i)+w(2)*DATAtest(2,i)+b;
RES=[RES sign(s)];
end
L1=length(find(RES(1:1:50)==1));
L2=length(find(RES(51:1:100)==- 1));
POS2=(L1+L2);
```

```
innerproduct.m
function [res]=innerproduct(x1,x2)
res=x1'*x2;
```

2.7 SVM with Soft Margin

In the classical SVM (with hard margin), we consider two parallel lines $\mathbf{w}^T\mathbf{x}_k + b = 1$ and $\mathbf{w}^T\mathbf{x}_k + b = -1$ such that $(\mathbf{w}^T\mathbf{x}_k + b)t_k > 1$ and the perpendicular distance between the line $\frac{2}{||\mathbf{w}||}$ is maximized. In this case, the vectors are not allowed between these parallel lines. This creates the problem of over fitting. Hence SVM with soft margin is used. In this case, a few vectors in between the parallel lines are allowed. This is done by introducing the positive quantity s_k (slack variable) associated with every training vector such that $(\mathbf{w}^T\mathbf{x}_k + b)t_k > 1 - s_k$. We can interpret that when $s_k = 0$, we impose the condition that the vector is lying in the correct side of the class (SVM with hard margin). We relax the condition if s_k is between 0 and 1. In this case, the corresponding vector $\mathbf{x}_k$ still lies in the correct side of the class region. If s_k is greater than 1, the vector lies in the wrong side of the class. We construct the objective function to minimize $\sum_{k=1}^{k=N} s_k$ with the constraints $s_k > 0$ for all k and $(\mathbf{w}^T\mathbf{x}_k + b)t_k \geq 1 - s_k$. This is solved using the Lagrangian equation constructed as follows:

$$B\sum_{k=1}^{k=N} s_k + \frac{||\mathbf{w}||^2}{2} - \sum_{u=1}^{u=N} a_u s_u - \sum_{v=1}^{v=N} b_v((\mathbf{w}^T\mathbf{x}_v + b)t_v - 1 + s_v) \tag{2.37}$$

B is fixed and is known as Box constant with the following KKT constraints:

$$a_k \geq 0 \tag{2.38}$$

$$b_k \geq 0 \tag{2.39}$$

$$((\mathbf{w}^T\mathbf{x}_k + b)t_k - 1 + s_k) \geq 0 \tag{2.40}$$

$$b_k((\mathbf{w}^T\mathbf{x}_k + b)t_k - 1 + s_k) = 0 \tag{2.41}$$

$$s_k \geq 0 \tag{2.42}$$

$$a_k s_k = 0 \tag{2.43}$$

Differentiating (2.37) with respect to $\mathbf{w}$ and b and equating to 0 to obtain the dual problem with the KKT conditions, we get the following:

$$\sum_{n=1}^{n=N} b_n - \frac{1}{2}\sum_{n=1}^{n=N}\sum_{m=1}^{m=N} b_n b_m t_n t_m \mathbf{x}_m^T \mathbf{x}_n \tag{2.44}$$

with the following constraints:

$$0 \leq b_n \leq B \tag{2.45}$$

$$\sum_{n=1}^{n=N} b_n t_n = 0 \tag{2.46}$$

$$b_n = B - a_n \tag{2.47}$$

Differentiating (2.37) with respect to a_r and equating to 0, we get the following:

$$[\mathbf{x}_1^T \mathbf{x}_r t_1 t_r \ \mathbf{x}_2^T \mathbf{x}_r t_2 t_r \cdots \mathbf{x}_N^T \mathbf{x}_r t_1 t_N][v_b] = 1 \tag{2.48}$$

for $r = 1 \cdots N$, where $\mathbf{v_b} = [b_1 \ b_2 \ \cdots b_N]^T$. The steps involved in getting the w and b in soft margin–based SVM are summarized as follows:

1. Solve b_k using (2.46) and (2.48).
2. Choose b_k that is positive and is bounded with B (i.e., $0 < b_k \leq B$).
3. Compute $\mathbf{w} = \sum_{m=1 \in M}^{m=N} b_m \mathbf{x}_m t_m$, where M is the set of support vectors corresponding to $b_k > 0$.
4. From (2.47), we understand that if $B > b_k > 0$, then $a_k > 0$. Equation (2.43) indicates that the corresponding $s_k = 0$. Hence (2.41) indicates $(\mathbf{w}^T \mathbf{x}_k + b)t_k - 1) = 0$. Thus b satisfies the equation

$$(\mathbf{w}^T \mathbf{x}_k + b)t_k - 1 = 0 \tag{2.49}$$

   for k that corresponds to $b_k > 0$.
5. The variable b is solved using (2.49) as $b = \frac{1}{N_s} \sum_{i=1, i \in M}^{i=N} (t_i - \mathbf{w}^T x_i)$, where N_s is the number of support vectors.

The soft-margin SVM is extended using the data that are mapped to the higher dimensional space using the "Kernel Trick." The steps involved in kernel-based soft-margin SVM are summarized as follows (Figs. 2.13, 2.14, and 2.15):

1. Formulate the equations

$$[k(\mathbf{x}_1, \mathbf{x}_r)t_1 t_r \ k(\mathbf{x}_2, \mathbf{x}_r)t_2 t_r \cdots k(\mathbf{x}_N, \mathbf{x}_r)][\mathbf{v}_{b\phi}] = 1 \tag{2.50}$$

   for $r = 1 \cdots N$, where $\mathbf{v_{b\phi}} = [b_{\phi 1} \ b_{\phi 2} \ \cdots b_{\phi N}]^T$
2. Solve for the elements of $\mathbf{v}_{b\phi}$, i.e., $b_{\phi 1}, b_{\phi 2}, ..., b_{\phi N}$ using (2.50) and $\sum_{n=1}^{n=N} t_n b_{\phi n} = 0$
3. Compute $\mathbf{w}_\phi = \sum_{m=1 \in M}^{m=N} b_m \phi(\mathbf{x}_m) t_m$, where M is the support vectors corresponding to $b_k > 0$. It is noted that $\phi(\mathbf{x_k})$, is not explicitly computed.
4. From $b_{\phi k} = B - a_{\phi k}$, we understand that if $B > b_{\phi k} > 0$, then $a_{\phi k} > 0$. Also $a_{\phi k} s_{\phi k} = 0$ indicates that the corresponding $s_{\phi k} = 0$. Hence $(\mathbf{w}_\phi^T \phi(\mathbf{x}_k) + b_\phi)t_k - 1) = 0$. Thus b_ϕ satisfies the equation

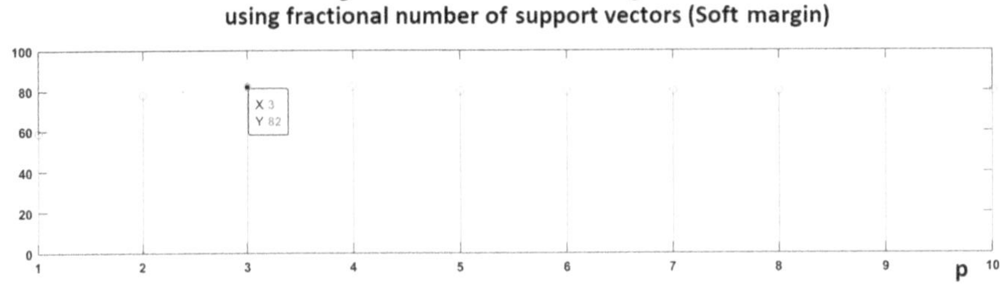

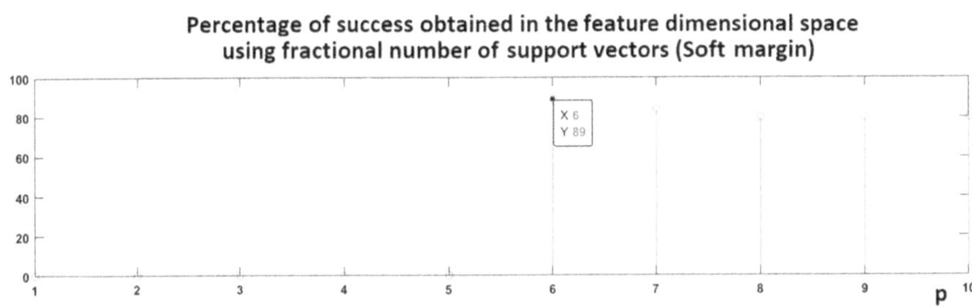

Fig. 2.13 Illustration of improvement achieved using fractional number of support vectors (number of support vectors $p\%$ of maximum value) of Lagrangian constant. Corresponding to Lagrangian constant greater than 0 and in soft margin–based SVM

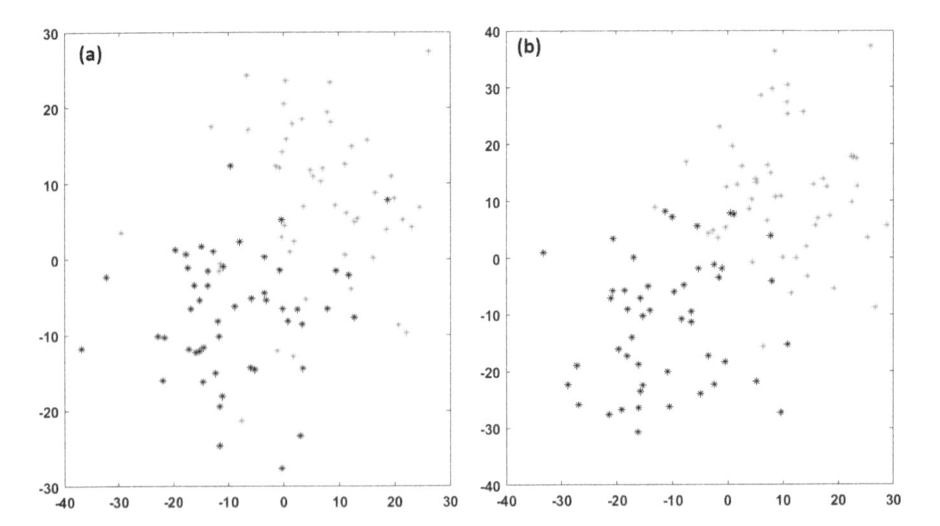

Fig. 2.14 (**a**) Training data with outliers used to train the soft margin–based SVM. (**b**) Corresponding testing data used

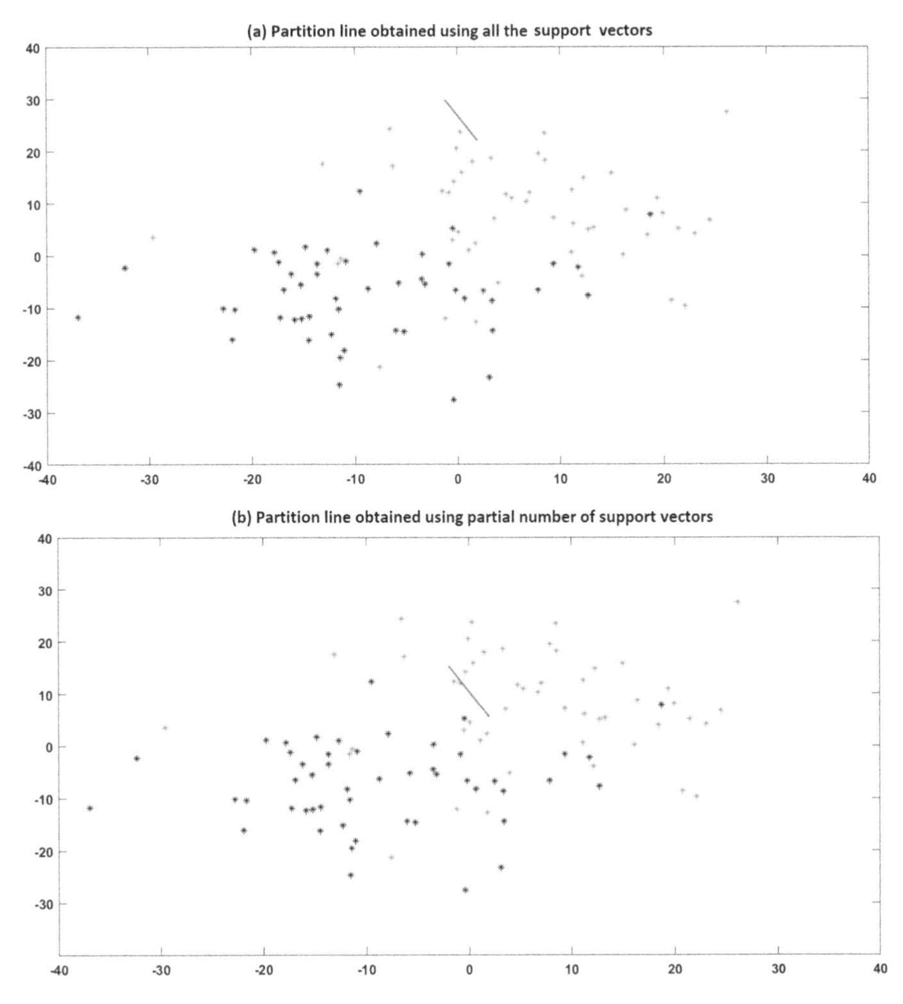

Fig. 2.15 (**a**) Partition line obtained using all the support vectors. (**b**) Partition line obtained using support vectors that has the Lagrangian constant lesser than 60% of the maximum value and greater than 0

$$(\mathbf{w}_{\phi}^{T}\mathbf{x}_{k} + b_{\phi})t_{k} - 1 = 0 \tag{2.51}$$

for k that corresponds to $b_{\phi k} > 0$

5. The variable $b_{\phi k}$ is solved using $b_{\phi k} = \frac{1}{N_s}\sum_{i=1, i \in M}^{i=N}(t_i - \mathbf{w}_{\phi}^{T}\phi(\mathbf{x}_i))$, where N_s is the number of support vectors.

softmargindemo.m

```
RES1=[];
for th=0.1:0.1:1
    RES1=[RES1 svm_softmargin1(th)];
end

RES2=[];
for th=0.1:0.1:1
    RES2=[RES2 svm_softmargin2(th)]
end

figure
subplot(2,1,1)
stem(RES1)
subplot(2,1,2)
stem(RES2)
```

svm_softmargin1.m

```
function [POS1]=svm_softmargin1(th)
load DATA
t=[ones(1,50) ones(1,50)*(-1)];
v=100;
figure(1)
subplot(1,2,1)
for i=1:1:50
    plot(DATAtrain(1,i),DATAtrain(2,i),'r*')
    hold on
end
for i=51:1:100
    plot(DATAtrain(1,i),DATAtrain(2,i),'k*')
    hold on
end

%Formulation of Gram-matrix
for i=1:1:100
    for j=1:1:100
        k(i,j)=gaussiankernel(DATAtrain(:,i),
        DATAtrain(:,j),v)*t(i)*t(j);
    end
end
M=[k;t];
W=pinv(M)*ones(101,1);
%Selection of Lagrangean constant greater
```

```
%than zero (Support vectors)
[p,q]=find((W>0)&(W<max(W)*th));
%b is obtained as follows.
temp=0;
for i=1:1:length(p)
   s1=t(q(i));
    s=0;
    for j=1:1:length(p)
    s=s+k(q(i),q(j))*t(q(j));
    end
    temp=temp+s1-s;
end
temp=temp/length(q);
%Check for the training set
RES=[];
for i=1:1:100
    s=0;
    for j=1:1:length(q)
s=s+gaussiankernel(DATAtrain(:,i),DATAtrain(:,q(j)),
v)*t(q(j));
    end
    s=s+temp;
RES=[RES sign(s)];
end
%Checking for the testing test
RES=[];
for i=1:1:100
    s=0;
    for j=1:1:length(q)
s=s+gaussiankernel(DATAtest(:,i),DATAtrain(:,q(j)),
v)*t(q(j))
    end
    s=s+temp;
RES=[RES sign(s)];
end
L1=length(find(RES(1:1:50)==1));
L2=length(find(RES(51:1:100)==- 1));
POS1=(L1+L2)
figure(1)
subplot(1,2,2)
for i=1:1:50
    plot(DATAtest(1,i),DATAtest(2,i),'r*')
    hold on
end
for i=51:1:100
```

```
    plot(DATAtest(1,i),DATAtest(2,i),'k*')
    hold on
end
```

svm_softmargin2.m

```
function [POS2]=svm_softmargin2(th)
load DATA
t=[ones(1,50) ones(1,50)*(-1)];
v=100;
figure
for i=1:1:50
    plot(DATAtrain(1,i),DATAtrain(2,i),'r*')
    hold on
end
for i=51:1:100
    plot(DATAtrain(1,i),DATAtrain(2,i),'k*')
    hold on
end
%Formulation of Gram-matrix
for i=1:1:100
    for j=1:1:100
        k(i,j)=innerproduct(DATAtrain(:,i),DATAtrain
        (:,j))*t(i)*t(j);
    end
end
M=[k;t];
W=pinv(M)*ones(101,1)
%Selection of Lagrangean constant greater
%than zero (Support vectors)
[p,q]=find((W>0)&(W<max(W)*th));
%b is obtained as follows.
temp=0;
for i=1:1:length(p)
    s1=t(q(i));
    s=0;
    for j=1:1:length(p)
    s=s+k(q(i),q(j))*t(q(j));
    end
    temp=temp+s1-s;
end
temp=temp/length(q);
s=0;
for i=1:1:length(q)
s=s+DATAtrain(:,q(i))*W(q(i))*t(q(i));
end
```

```
w=s;
b=0;
for i=1:1:length(p)
    b=b+t(q(i))-w(1)*DATAtrain(1,q(i))+w(2)*DATAtrain
    (2,q(i))
end
b=b/length(q);
if(isempty(p)==0)
x=-2:0.1:2;
y=(ones(1,length(x))*(-b)-w(1)*x)/w(2);
hold on
plot(x,y,'-')
%Check for the training set
RES=[];
for i=1:1:100
    s=w(1)*DATAtrain(1,i)+w(2)*DATAtrain(2,i)+b;
    RES=[RES sign(s)];
end
%Checking for the testing test
RES=[];
for i=1:1:100
s=w(1)*DATAtest(1,i)+w(2)*DATAtest(2,i)+b;
RES=[RES sign(s)];
end
L1=length(find(RES(1:1:50)==1));
L2=length(find(RES(51:1:100)==- 1));
POS2=(L1+L2);
end
if(isempty(p)==1)
POS2=0;
end
```

2.8 SVM for Regression

We consider the problem of regression with the target value as t_n corresponding to the input vector $\mathbf{x}_n$ using $y_n = \mathbf{w}_T\mathbf{x}_n = \mathbf{b}$. The vector $\mathbf{w}$ is obtained by minimizing $\sum_{k=1}^{k=N}(t_k - y_k)^2$, subject to the constraint $\mathbf{w}^T\mathbf{w}$ is minimized. The Largrangian function is structured as follows:

$$J = \sum_{k=1}^{k=N}(t_k - y_k)^2 - \frac{\lambda}{2}\mathbf{w}^T\mathbf{w} \tag{2.52}$$

This is solved using Support Vector Machine (SVM). The objective function is approximated as follows. Let $z_k = t_k - y_k$.

1. If $|z_k| = |y_k - t_k| < \delta$, the cost associated with the objective function is 0. Also if $|z_k = y_k - t_k| > \delta$, the cost associated with that is $|z_k| = |y_k - t_k|$.
2. Consider the margin $z_k + \lambda$ and $z_k - \lambda$. We need to find the optimal value for the vector $\mathbf{w}$ and b such that the vectors are inside the margin. If the vectors are outside the margin, we consider it as an error.
3. We allow a few vectors outside this margin. These are controlled by slack variables (non-zero) assigned to each vector such that $\mathbf{w}^T\mathbf{x}_n + b - t_n \leq \delta + \alpha_n$ and $\mathbf{w}^T\mathbf{x}_n + b - t_n \geq -\delta - \beta_n$.
4. Thus problem is structured as minimization problem as follows:

$$J = \sum_{k=1}^{k=N}(\alpha_k + \beta_k) + \frac{\lambda}{2}||\mathbf{w}||^2 \tag{2.53}$$

$$J = C\sum_{k=1}^{k=N}(\alpha_k + \beta_k) + ||\mathbf{w}||^2 \tag{2.54}$$

5. The constraints are given as follows:

$$\mathbf{w}^T x_n + b - t_n \leq \delta + \alpha_n \tag{2.55}$$

$$\mathbf{w}^T x_n + b - t_n \geq -\delta - \beta_n \tag{2.56}$$

$$\alpha_n > 0 \tag{2.57}$$

$$\beta_n > 0 \tag{2.58}$$

6. The Lagrangian equation is structured as follows:

$$C\sum_{k=1}^{k=N}(\alpha_k + \beta_k) + ||\mathbf{w}||^2 - \sum_{n=1}^{n=N} a_n(t_n - \mathbf{w}^T\mathbf{x}_n - b + \delta + \alpha_n) + \sum_{n=1}^{n=N} b_n(t_n - \mathbf{w}^T\mathbf{x}_n - b - \delta - \alpha_n) - \sum_{n=1}^{n=N} c_n\alpha_n - \sum_{n=1}^{n=N} d_n\beta_n \tag{2.59}$$

7. Differentiating the Lagrangian equation, we get the following:

$$\mathbf{w} = \sum_{k=1}^{k=N}(a_k - b_k)\mathbf{x}_k \tag{2.60}$$

$$\sum_{k=1, k\in M}^{k=N} a_k = \sum_{k=1}^{k=N} b_k \tag{2.61}$$

$$a_k = C - c_k \tag{2.62}$$

$$b_k = C - d_k \tag{2.63}$$

where M corresponds to the set of support vectors.

8. The dual maximization problem is obtained as follows:

$$-\frac{1}{2}\sum_{n=1}^{n=N}\sum_{m=1}^{m=N}(a_n - b_n)(a_m - b_m)\mathbf{x}_n^T\mathbf{x}_m - \delta\sum_{n=1}^{n=N}(a_n + b_n) + \sum_{n=1}^{n=N}(a_n - b_n)t_n \tag{2.64}$$

The steps involved in obtaining the regression curve using SVM are summarized as follows (Fig. 2.16):

1. Differentiating (2.64) with respect to a_1 and equating to 0, we get the following:

$$[\mathbf{x}_1^T\mathbf{x}_1 a_1 + \mathbf{x}_1^T\mathbf{x}_2 a_2 + \cdots x_1^T\mathbf{x}_N a_N - \mathbf{x}_1^T\mathbf{x}_2 b_2 - \mathbf{x}_1^T\mathbf{x}_3 b_3 \cdots - \mathbf{x}_1^T x_N b_N] = t_1 - \delta \tag{2.65}$$

   From (2.65), it is observed that the coefficient of b_1 is 0 in the expression.

2. Similarly, differentiating (2.64) with respect to b_1 and equating to 0, we get the following:

$$[\mathbf{x}_1^T\mathbf{x}_2 a_2 + \cdots \mathbf{x}_1^T\mathbf{x}_N a_N - \mathbf{x}_1^T\mathbf{x}_1 b_1 - \mathbf{x}_1^T\mathbf{x}_2 b_2 - \mathbf{x}_1^T\mathbf{x}_3 b_3 \cdots - \mathbf{x}_1^T\mathbf{x}_N b_N] = t_n + \delta \tag{2.66}$$

   In this case, it is seen that the coefficient of a_1 is 0.

3. Differentiating with respect to $a_1, a_2, ..., a_N, b_1, b_2, ..., b_N$ and equating to 0, we get $2N$ equations.

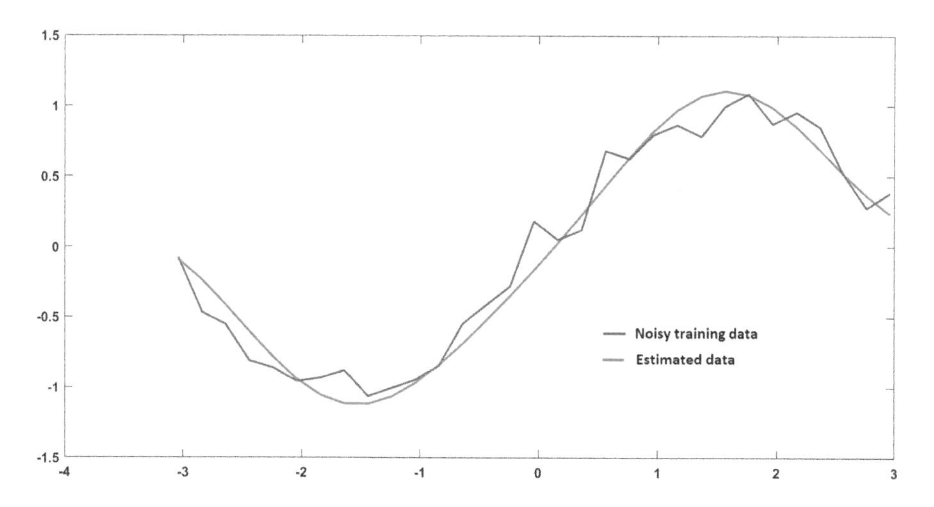

Fig. 2.16 Demonstration of Support Vector Machine used as regression model

4. Combining all the equations, we get the following ($N = 3$ is chosen for the illustration):

$$\begin{pmatrix} \mathbf{x}_1^T\mathbf{x}_1 & \mathbf{x}_1^T\mathbf{x}_2 & \mathbf{x}_1^T\mathbf{x}_3 & 0 & -\mathbf{x}_1^T\mathbf{x}_2 & -\mathbf{x}_1^T\mathbf{x}_3 \\ \mathbf{x}_2^T\mathbf{x}_1 & \mathbf{x}_2^T\mathbf{x}_2 & \mathbf{x}_2^T\mathbf{x}_3 & -\mathbf{x}_2^T\mathbf{x}_1 & 0 & -\mathbf{x}_2^T\mathbf{x}_3 \\ \mathbf{x}_3^T\mathbf{x}_1 & \mathbf{x}_3^T\mathbf{x}_2 & \mathbf{x}_3^T\mathbf{x}_3 & -\mathbf{x}_3^T\mathbf{x}_1 & -\mathbf{x}_3^T\mathbf{x}_2 & 0 \\ 0 & \mathbf{x}_1^T\mathbf{x}_2 & \mathbf{x}_1^T\mathbf{x}_3 & -\mathbf{x}_1^T\mathbf{x}_1 & -\mathbf{x}_1^T\mathbf{x}_2 & -\mathbf{x}_1^T\mathbf{x}_3 \\ \mathbf{x}_2^T\mathbf{x}_1 & 0 & \mathbf{x}_2^T\mathbf{x}_3 & -\mathbf{x}_2^T\mathbf{x}_1 & -\mathbf{x}_2^T\mathbf{x}_2 & -\mathbf{x}_2^T\mathbf{x}_3 \\ \mathbf{x}_3^T\mathbf{x}_1 & \mathbf{x}_3^T\mathbf{x}_2 & 0 & -\mathbf{x}_3^T\mathbf{x}_1 & -\mathbf{x}_3^T\mathbf{x}_2 & -\mathbf{x}_3^T\mathbf{x}_3 \end{pmatrix} \begin{pmatrix} a_1 \\ a_2 \\ a_3 \\ b_1 \\ b_2 \\ b_3 \end{pmatrix} = \begin{pmatrix} t_1 - \delta \\ t_2 - \delta \\ t_3 - \delta \\ -t_1 - \delta \\ -t_2 - \delta \\ -t_3 - \delta \end{pmatrix}$$

5. Solving for a_k's and b_k's and choosing the values greater than 0.
6. Computing w using (2.60). As the solution satisfies the constraints (KKT), we have the following:

$$a_n(\mathbf{w}^T\mathbf{x}_n + b - t_n - \delta - \alpha_n) = 0 \tag{2.67}$$

$$b_n(\mathbf{w}^T\mathbf{x}_n + b - t_n + \delta + \beta_n) = 0 \tag{2.68}$$

$$(C - a_n)\alpha_n = 0 \tag{2.69}$$

$$(C - b_n)\beta_n = 0 \tag{2.70}$$

7. For $a_n > 0$ and less than C, the optimal point that maximizes the dual problem satisfies the condition $\alpha_n = 0$ and $\beta_n = 0$.
8. Thus b is obtained using the vector corresponding to $b_n > 0$ and $a_n > 0$.

$$\mathbf{w}^T\mathbf{x}_n + b - t_n - \delta = 0 \tag{2.71}$$

$$\Rightarrow b = \delta + t_n - \mathbf{w}^T\mathbf{x}_n\mathbf{w}^T\mathbf{x}_n + b - t_n + \delta = 0 \tag{2.72}$$

$$\Rightarrow b = -\delta + t_n - \mathbf{w}^T\mathbf{x}_n \tag{2.73}$$

9. Combining both the equations, we get $b = t_n - \mathbf{w}^T\mathbf{x}_n$. Taking the average over all the vectors, we get the following:

$$b = \frac{1}{n(M)} \sum_{k=1,\in M} k = N(t_k - \mathbf{w}^T\mathbf{x}_k) \tag{2.74}$$

where $n(M)$ is the number of support vectors (corresponding to $a_k > 0$ and $b_k > 0$).

svmforregression.m

```
close all
clear all
%SVM for regression
x=-3.14:0.1:3.14;
y=sin(x)+0.1*randn(1,length(x));
```

```
%%%%%%%%%%%%%%%%%%%%%%%%%%%%%%%%%
TRAINDATAx=x(1:2:length(x));
TRAINDATAy=y(1:2:length(y));
TESTDATAx=x(2:2:length(x));
TESTDATAy=y(2:2:length(y));
%Formulation of matrix to compute a's and b's
sigma=0.9;
M=[];
PART1=[];
PART2=[];
for i=1:1:length(TRAINDATAx)
    temp=[];
    for j=1:1:length(TRAINDATAx)
temp=[temp gaussiankernel(TRAINDATAx(i),TRAINDATAx(j),
sigma)] ;
    end
    temp1=temp;
    temp1(i)=0;
    PART1=[PART1;temp -1*temp1];
    PART2=[PART2;-1*temp1 temp];
end
M=[PART1;PART2];
n=size(TRAINDATAx,2);
M=[M;ones(1,n) -1*ones(1,n)];
v=[TRAINDATAy-0.01*ones(1,n) -ones(1,n)*0.01-
TRAINDATAy 0];
vab=pinv(M)*v';
vab1=reshape(vab,length(vab)/2,2);
vab1=vab1';
POS=[];
for i=1:1:size(vab1,2)
if((vab1(1,i)>0)|((vab1(2,i)>0)))
    POS=[POS i];
 end
end
POS
COL=vab1(:,POS);
s=0;
for i=1:1:length(COL)
    temp=0;
    for j=1:1:length(COL)
temp=temp+(COL(1,j)-COL(2,j))*gaussiankernel
(TRAINDATAx(POS(j)),TRAINDATAx(POS(i)),sigma);
    end
    s=s+(TRAINDATAy(POS(i))-temp);
```

```
end
s=s/(length(COL));
ESTIMATEy=[];
for i=1:1:length(TESTDATAx)
    temp=0;
    for j=1:1:length(COL)
temp=temp+(COL(1,j)-COL(2,j))*gaussiankernel
(TRAINDATAx(POS(j)),TESTDATAx(i),sigma);
    end
    ESTIMATEy=[ESTIMATEy temp+s];
end
figure
plot(TESTDATAx,ESTIMATEy,'r')
hold on
plot(TESTDATAx,TESTDATAy,'b')
```

2.9 RSVM for Classification

The steps involved in probabilistic approach (logistic regression) in classification is summarized as follows:

1. Let the prior probability of the kth class index be represented as $p(c_k)$. The a posteriori probability of the class index is represented as $p(c_k/\mathbf{x})$, and the emission probability or generating probability of random vector $\mathbf{x}$ given the class label is given as $p(\mathbf{x}/c_k)$.
2. In the logistic regression model, $p(c_k/\mathbf{x})$ is modeled directly as soft-max function, given as follows:

$$y_{nk} = p(c_k/\mathbf{x}_n) = \frac{e^{w_k^T \phi(\mathbf{x})}}{\sum_{i=1}^{i=r} e^{w_i^T \phi(\mathbf{x}_n)}} \tag{2.75}$$

   where r is the total number of classes.
3. Given N training data, the unknown $\mathbf{w}$'s are obtained by maximizing the likelihood function $p(t_1 t_2 \cdots t_N/\mathbf{w}) = \Pi_{i=1}^{i=N} \Pi_{j=1}^{j=r} (y_{nk})^{t_{nk}}$, where t_{nk} takes the value 1 if the nth vector belongs to the kth class. Otherwise it takes the value 0.
4. Equivalently $\mathbf{w}$'s are obtained by minimizing $-log$ of the likelihood function, which is given as follows:

$$J = -\sum_{i=1}^{i=N} \sum_{j=1}^{j=r} t_{nk} log(y_{nk}) \tag{2.76}$$

5. This can be solved using gradient method or Newton–Raphson method as follows:
   Gradient method

$$\mathbf{w}_j(t+1) = \mathbf{w}_j(t) - \eta \nabla_j J \tag{2.77}$$

   Newton–Raphson method

$$\mathbf{w}_j(t+1) = \mathbf{w}_j(t) - \eta H_j^{-1} \nabla_j J \tag{2.78}$$

2.9.1 Bayesian Approach to Estimate w

In this technique, the prior density of the random vector $\mathbf{w}$ is Gaussian distributed with mean zero and co-variance matrix $\mathbf{A}$, which is diagonal matrix with the ith diagonal element is $\frac{1}{\alpha_i}$. The posterior density function of the random vector $p(\mathbf{w}/t, \alpha)$ is approximated using Gaussian density function as follows:

$$J_1 = ln(p(\mathbf{w}/t, \alpha)) = ln(p(t/\mathbf{w})p(\mathbf{w}/\alpha)) - ln(p(t/\alpha)) \tag{2.79}$$

$$\sum_{n=1}^{n=N} (t_n ln(y_n) + (1-t_n) ln(1-y_n) - \frac{1}{2} w^T A w + constant \tag{2.80}$$

We approximate the posterior density function as the Gaussian density function at the peak of the function J_1. This is obtained by differentiating J_1 with respect to $\mathbf{w}$, equating to 0 and solving for $\mathbf{w}$ as $\mathbf{w}_{mode}$. At this value, the posterior density function $p(\mathbf{w}/t)$ is approximated as Gaussian with mean vector $\mathbf{w}_{mode}$ and the co-variance matrix as the inverse of the Negative Hessian matrix of $J_1 = ln(p(\mathbf{w}/t, \boldsymbol{\alpha}))$. This technique is based on the Laplacian approximation of the pdf at the arbitrarily selected point of the pdf under consideration. The gradient of the objective function J_1 and the co-variance matrix computed of the function J_1 is obtained as follows:

$$\nabla J_1 = \Phi^T(t-y) - \mathbf{A}\mathbf{w} \tag{2.81}$$

$$\nabla\nabla J_1 = -(\Phi^T \mathbf{B} \Phi + \mathbf{A}) \tag{2.82}$$

where $\mathbf{B}$ is the diagonal matrix with diagonal elements $y_{nj}(1-y_{nj})$. Thus mode and the Hessian at the mode are obtained as follows:

$$\mathbf{w}_{mode} = \mathbf{A}^{-1}\Phi^T(t-y) \tag{2.83}$$

$$\mathbf{C} = (\Phi^T \mathbf{B} \Phi + \mathbf{A})^{-1} \tag{2.84}$$

We know

$$p(\mathbf{w}/t, \boldsymbol{\alpha}) = \frac{p(t/\mathbf{w})p(\mathbf{w}/\boldsymbol{\alpha})}{p(t/\boldsymbol{\alpha})}$$

$$p(\mathbf{w}/t, \boldsymbol{\alpha})p(t/\boldsymbol{\alpha}) = p(t/\mathbf{w})p(\mathbf{w}/\boldsymbol{\alpha})$$

It is understood that $p(t/\mathbf{w})p(\mathbf{w}/\alpha)$ is approximated as Gaussian distributed at $\mathbf{w}_{mode}$ as $p(\mathbf{w}/t,\alpha)$ is assumed as Gaussian distributed at $\mathbf{w}_{mode}$. Taking marginal integral over w, we get the following:

$$\int p(\mathbf{w}/t, \boldsymbol{\alpha})p(t/\boldsymbol{\alpha})d\mathbf{w} = \int p(t/\mathbf{w})p(\mathbf{w}/\boldsymbol{\alpha})d\mathbf{w} \quad (2.85)$$

$$\Rightarrow \int p(t/\boldsymbol{\alpha})d\mathbf{w} = \int p(t/\mathbf{w})p(\mathbf{w}/\boldsymbol{\alpha})d\mathbf{w} \quad (2.86)$$

Using Laplace approximation, we represent $\int p(t/\alpha)dw = p(t/\mathbf{w}_{mode})$ $p(\mathbf{w}_{mode}/\alpha)(2\Pi)^{M/2}|C|^{1/2}$. We optimize α_i by maximizing the likelihood function $\int p(t/\alpha)dw$. Thus the α_i is updated as $\gamma_i/\mathbf{w}^2_{mode_i}$. Note that α_i is the ith element of the vector $\boldsymbol{\alpha}$, where $\gamma_i = 1 - \alpha_i C(i,i)$ and $\mathbf{w}_{mode_i}$ is the ith element of the vector $\mathbf{w}_{\mathbf{mode}}$ (Figs. 2.17, 2.18, and 2.19). The steps involved in RVM-based classification (for two class) are summarized as follows. Note that in the case of two class problem, the negative logarithm of the likelihood function is given as follows: $-\sum t_k ln(y_k - \sum(1-t_k)ln(1-y_k)$

1. Initialize $\mathbf{w}$ vector and $\boldsymbol{\alpha}$.
2. Obtain y_{nk} using (2.76) for all the training vectors.
3. Formulate matrix $\mathbf{A}$ with i^{th} diagonal element is $\frac{1}{\alpha_i}$ using (2.84).
4. Compute the matrix $\mathbf{C}$ using (2.84) and $\mathrm{w_{mode}}$ using (2.83).

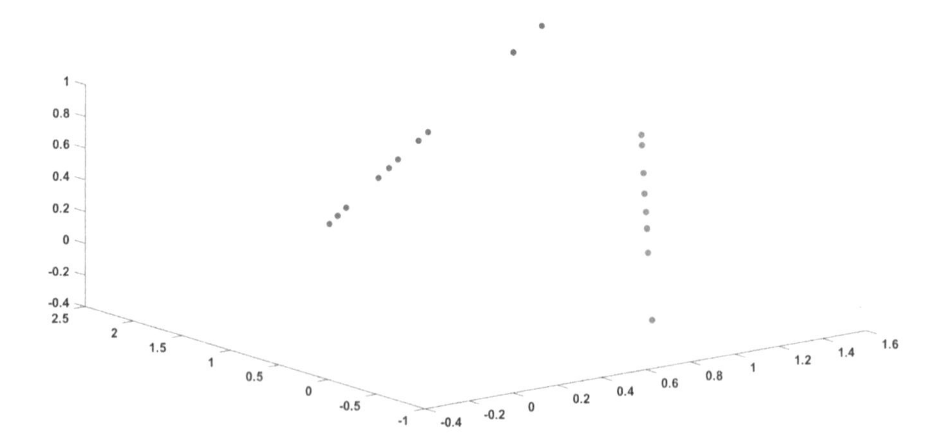

Fig. 2.17 Two class data subjected to RSVM-based classification

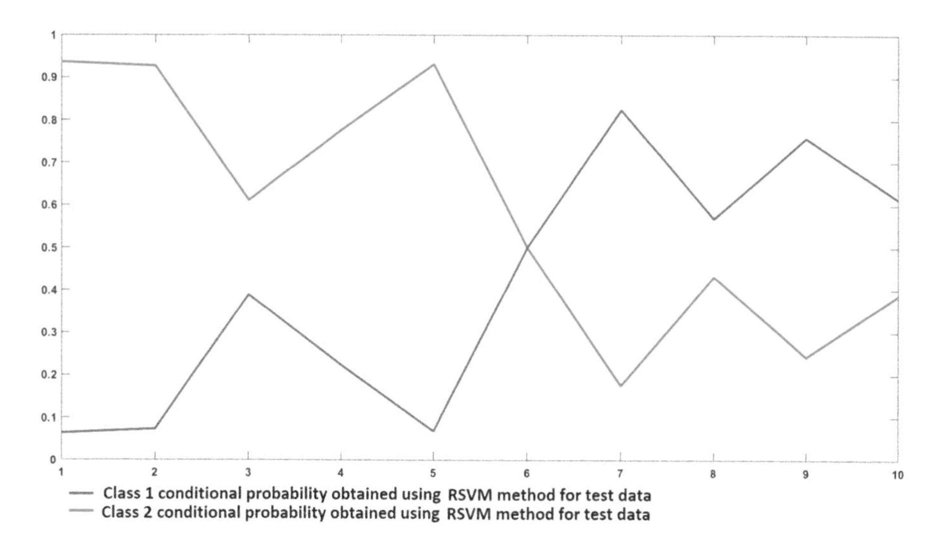

Fig. 2.18 Illustration of class conditional probabilities obtained using the test data

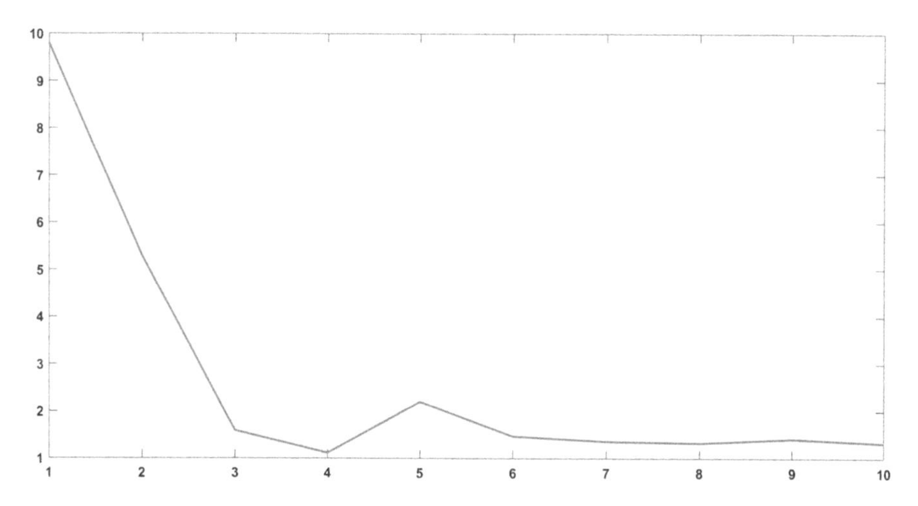

Fig. 2.19 Illustration of convergence of the cross-entropy obtained using RSVM

5. Compute $\gamma_i = 1 - \alpha_i C_{ii}$.
6. Compute $\alpha_i = \gamma_i \mathbf{w}_{mode}$.
7. Repeat the steps (2)–(5) until J is converged.

RVMforclassification.m

```
close all
clear all
%RVM sparsity for classification
%Multi-class problem
```

```
m1=[1 0 0]';
m2=[0 1 0]';
m1=rand(3,1)+m1;
m2=rand(3,1)+m2;
N1=10; N2=10;
POSTRAIN=[];
POSTEST=[];
POSVAL=[];
temp1=rand(length(m1),1);
C1=temp1*temp1'*0.2;
temp2=rand(length(m2),1);
C2=temp2*temp2'*0.2;
[X1,C1est]=genrandn(m1,C1,N1);
[X2,C2est]=genrandn(m2,C2,N2);
DATA=[X1 X2];
figure(1)
plot3(X1(1,:),X1(2,:),X1(3,:),'r*')
hold on
plot3(X2(1,:),X2(2,:),X2(3,:),'b*')
[V,M]=kmeans(DATA',4);
B1=[]; B2=[]; B3=[];  B4=[];
sigma=0.5;
for i=1:1:size(DATA,2)
B1=[B1 gbf(DATA(:,i)',M(1,:),sigma)];
B2=[B2 gbf(DATA(:,i)',M(2,:),sigma)];
B3=[B3 gbf(DATA(:,i)',M(3,:),sigma)];
B4=[B4 gbf(DATA(:,i)',M(4,:),sigma)];
end
%Formulation of matrix
MAT=[ones(20,1) B1' B2' B3' B4'];
TRAIN=[MAT(1:1:5,:); MAT(11:1:15,:)];
TEST=[MAT(6:1:10,:); MAT(16:1:20,:)];
%Initialize A matrix
TRAININDEX=[ones(1,5) zeros(1,5);]
TESTINDEX=[ones(1,5) zeros(1,5);]
%Initialize weights
%Computation of ynk (nth vector in the kth class)
%Initializing W and alpha
load INITIALIZE
ERROR=[];
for iteration=1:1:10
%Soft max computation for the individual classses
P1=(1./(1+exp(-TRAIN*W)));
P2=1-P1;
%Error function computation
```

```
P=[P1 P2]
error=-1*sum(sum([ones(1,5) zeros(1,5);zeros(1,5)
ones(1,5)]'.*log(P)));
ERROR=[ERROR error];
%Updating the weights are obtained as follows.
%Initialize alphai's matrix
A=diag(alpha);
W=(inv(A))*TRAIN'*(TRAININDEX'-P1);
B=diag([P1.*(1-P1)]);
C=pinv(TRAIN'*B*TRAIN+A);
%Updata W
for i=1:1:5
    gamma(i)=1-alpha(i)*C(i,i);
end
alpha=(gamma)./(W'.^2);
end
%Checking with test data
%Soft max computation for the individual classses
P1=(1./(1+exp(-TEST*W)));
P2=1-P1;
%Error function computation
P=[P1 P2];
[u,v]=min(P');
REF=[ones(1,5) ones(1,5)*2];
POS=length((find(REF-v)==0))*10
figure(2)
plot(P(:,1),'r')
hold on
plot(P(:,2),'b')
figure(3)
plot(ERROR)
```

Chapter 3
Regression Techniques

3.1 Linear Regression

3.1.1 The Random Variable X Influencing Y

Let the prior probability density function of the random variable X be represented as $f_X(x)$ and the a posteriori density function of the random variable x given $Y = y$ is given as $f_{X/Y=y}(x)$. The likelihood function is given as $g_{Y/X=x}(y)$. Let the estimate of the random variable X based on the observation of the random variable $Y = y$ is represented as $X\hat{(y)}$. The cost involved in assigning X with $X\hat{(y)}$ is given as $J = C(x, x\hat{(y)})$. We would like to get the solution that minimizes $E(C(x, x\hat{(y)}))$.

$$E(C(x, x\hat{(y)})) = \int_{-\infty}^{\infty} C(x, x\hat{(y)}) f_{XY}(x, y) dx dy$$

$$= \int_{-\infty}^{\infty} \int_{-\infty}^{\infty} C(x, x\hat{(y)}) f_{X/Y=y}(x) dx f_Y(y) dy$$

$$= E_Y(E(C(x, x\hat{(y)}))/Y = y)$$

This is equivalent to minimizing $E(C(x, x\hat{(y)}))/Y = y)$ for every values of $Y = y$.

1. Minimum Mean Square Estimation: Estimating x to obtain $x\hat{(y)}$ by minimizing

$$E(C(x, x\hat{(y)}))/Y = y) = E((x - x\hat{(y)})^2/Y = y) \tag{3.1}$$

© Springer Nature Switzerland AG 2020

E. S. Gopi, *Pattern Recognition and Computational Intelligence Techniques Using Matlab*, Transactions on Computational Science and Computational Intelligence,
https://doi.org/10.1007/978-3-030-22273-4_3

Differentiating (3.1) with respect to $x\hat{(y)}$, equating to zero, and solving for $x\hat{(y)}$ are given as follows:

$$\int_{-\infty}^{\infty} 2(x - x\hat{(y)}) f_{X/Y=y}(x)dx = 0 \tag{3.2}$$

$$= \int_{-\infty}^{\infty} 2x\hat{(y)} f_{X/Y=y}(x)dx = \int_{-\infty}^{\infty} x f_{X/Y=y}(x)dx \tag{3.3}$$

$$= x\hat{(y)}_{MMSE} = \int_{-\infty}^{\infty} x f_{X/Y=y}(x)dx \tag{3.4}$$

This estimation is called Minimum Mean Square Estimation (MMSE).

2. Minimum Mean Absolute Estimation (MMAE): Estimating X to obtain $X\hat{(y)}$ by minimizing

$$E(C(x, x\hat{(y)})/Y = y) = E(|x - x\hat{(y)}|/Y = y) \tag{3.5}$$

Differentiating (3.5) with respect to $x\hat{(y)}$, equating to zero, and solving for $X\hat{(y)}$ are given as follows:

$$\int_{-\infty}^{\infty} |x - x\hat{(y)}| f_{X/Y=y}(x)dx \tag{3.6}$$

$$= \int_{-\infty}^{X\hat{(y)}} (x\hat{(y)} - x) f_{X/Y=y}(x)dx + \int_{X\hat{(y)}}^{\infty} (x - x\hat{(y)}) f_{X/Y=y}(x)dx \tag{3.7}$$

$$\int_{X}^{x\hat{(y)}} -f_{X/Y=y}(x)dx + \int_{x\hat{(y)}}^{\infty} f_{X/Y=y}(x)dx = 0 \tag{3.8}$$

$$\Rightarrow \int_{-\infty}^{x\hat{(y)}_{MMAE}} f_{X/Y=y}(x)dx = \int_{x\hat{(y)}_{MMAE}}^{\infty} f_{X/Y=y}(x)dx = 0 \tag{3.9}$$

This estimation is called Minimum Mean Square Estimation (MMSE).

3. Consider the cost $C(x, x\hat{(y)}) = \frac{1}{\Delta}$, if $|x - x\hat{(y)}| \geq \frac{\Delta}{2}$, 0, otherwise.

$$E(C(x, x\hat{(y)})/Y = y) = \frac{P(|x - x\hat{(y)}| \geq \frac{\Delta}{2}/Y = y)}{\Delta} \tag{3.10}$$

$$= \frac{1 - P(|x - x\hat{(y)}| \leq \frac{\Delta}{2}/Y = y)}{\Delta} \tag{3.11}$$

$$= \frac{1 - P(x\hat{(y)} - \frac{\Delta}{2} \leq X \leq x\hat{(y)} + \frac{\Delta}{2}/Y = y)}{\Delta} \tag{3.12}$$

$$= \frac{1 - f(X\hat{(y)}/Y = y)\Delta}{\Delta} \tag{3.13}$$

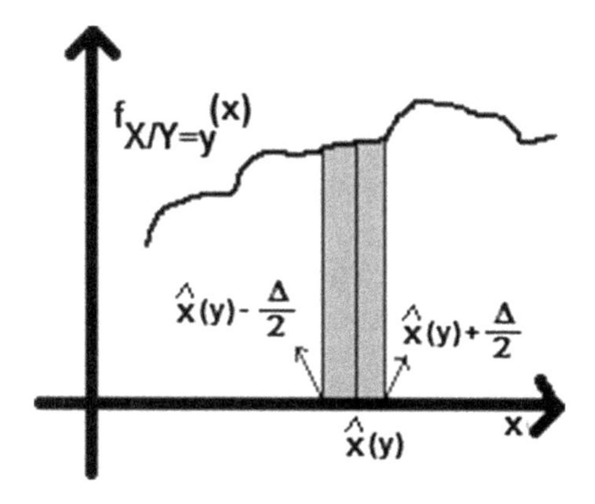

Fig. 3.1 Illustration of maximum a posteriori probability estimate

We would like to minimize $E(C(x, x\hat{(y)})/Y = y) = \frac{1-f(x\hat{(y)}/Y=y)\Delta}{\Delta}$ with Δ tends to zero (refer Fig. 3.1).

$$Min \quad Lt\Delta \to 0 \quad \frac{1 - f(x\hat{(y)})\Delta/Y = y}{\Delta} = Min - f(x\hat{(y)}/Y = y) \tag{3.14}$$

$$\Rightarrow Max \quad f(x\hat{(y)}/Y = y) \tag{3.15}$$

Thus $x\hat{(y)}$ is obtained by maximizing the a posteriori density function $f_{X/Y=y}(x)$. This estimation is known as Maximum A Posteriori Probability (MAP) estimate.

3.2 Parametric Approach

Let the outcome of the random variables x_i, for $i = 1 \cdots n$, influence the outcome of the random variables y_j, for $j = 1 \cdots m$. Let

$$\mathbf{x} = [x_1 \ x_2 \ \cdots x_n]^T$$

$$\mathbf{y} = [y_1 \ y_2 \ \cdots y_m]^T$$

We are in need of estimating the outcome of the random vector **y**, given the outcome of the random vector **x**.

This is obtained by constructing the relationship between **y** and **x** using the linear relationship as $\mathbf{y} = \mathbf{W}\mathbf{f}$, where $\mathbf{f} = [f_1(\mathbf{x}) \ f_2(\mathbf{x}) \cdots f_r(\mathbf{x})]^T$. The matrix **W** is given as follows:

$$\mathbf{W} = \begin{pmatrix} w_{11} & w_{12} & \cdots & w_{1r} \\ w_{21} & w_{22} & \cdots & w_{2r} \\ \cdots & \cdots & \cdots & \cdots \\ w_{m1} & w_{m2} & \cdots & w_{mr} \end{pmatrix}.$$

Regression involves estimating the values for the matrix $\mathbf{W}$ such that the error involved in assigning the estimated values with the original values is minimized. MMSE estimate of $\mathbf{W}$ is obtained as the conditional mean of the posterior density function $f_{\mathbf{W}/Y=y}$.

Let us consider that the data collected for the input random vector $\mathbf{x}$ are given as $\mathbf{x_1}, \mathbf{x_2}, ..., \mathbf{x_d}$ and the noisy version of the corresponding output vector is collected as $\mathbf{t_1}, \mathbf{t_2}, ..., \mathbf{t_d}$, where $\mathbf{t_i} = \mathbf{x_i} + \boldsymbol{\epsilon_i}$, where $\boldsymbol{\epsilon_i}$ is the additive Gaussian noise vector with mean zero and co-variance matrix $\boldsymbol{\beta}^{-1}\mathbf{I}$ and $\mathbf{I}$ is the identity matrix. Least square technique to estimate the value of the matrix $\mathbf{W}$ is obtained by minimizing $|\mathbf{T} - \mathbf{F}\mathbf{W}^T|^2$. It is noted that the size of the matrices $\mathbf{W}$, $\mathbf{F}$, and $\mathbf{T}$ is $r \times m$, $d \times m$, and $d \times n$, respectively. It is noted that d is the number of data used and r is the number of basis function used to estimate $\mathbf{W}$. Using pseudo-inverse, the matrix $\mathbf{W}$ is estimated as follows:

$$\widehat{\mathbf{W}}^T = (\mathbf{F}^T\mathbf{F})^{-1}\mathbf{F}^T\mathbf{T} \tag{3.16}$$

3.2.1 Illustrations on Parametric Method

The data is generated using the following. $\mathbf{x}$ is the input vector with size 10×1, which is uniformly distributed between -1 and 1. Thousand such data are generated. The output is obtained as follows:

$$\mathbf{h_1} = \mathbf{W_1}\mathbf{x} \tag{3.17}$$

$$\mathbf{h_2} = \frac{1}{1+e^{-\mathbf{h_1}}} \tag{3.18}$$

$$\mathbf{h_3} = \mathbf{W_2}\mathbf{h_2} \tag{3.19}$$

$$\mathbf{y} = \frac{1}{1+e^{-\mathbf{h_3}}} \tag{3.20}$$

$$\mathbf{t} = \mathbf{y} + \boldsymbol{\epsilon} \tag{3.21}$$

where $\mathbf{W_1}$ and $\mathbf{W_2}$ are the outcome of the random matrix with size 5×10 (uniformly distributed between 0 and 1) and 5×3, respectively. The corresponding output vector $\mathbf{y}$ is added with the outcome of the Gaussian random vector $\boldsymbol{\epsilon}$ (noise model). It is seen that the size of the random vector $\mathbf{y}$ and $\boldsymbol{\epsilon}$ is identical as 3×1. The co-variance matrix of the random vector $\boldsymbol{\epsilon}$ is given as $\sqrt{(0.2)}\mathbf{I}$. This implies that 0.2 is the variance of the individual element of the random vector $\boldsymbol{\epsilon}$ and the elements of the random vector are statistically independent. The regression technique using parametric method is illustrated in Figs. 3.2, 3.3, and 3.4.

Out of 1000, 700 training data are used to estimate the matrix $\mathbf{W}$. The remaining 300 data are used as the testing data and are compared with actual value $\mathbf{y}$ (not $\mathbf{t}$). The number of basis functions r used in this experiment is 4. The first basis function

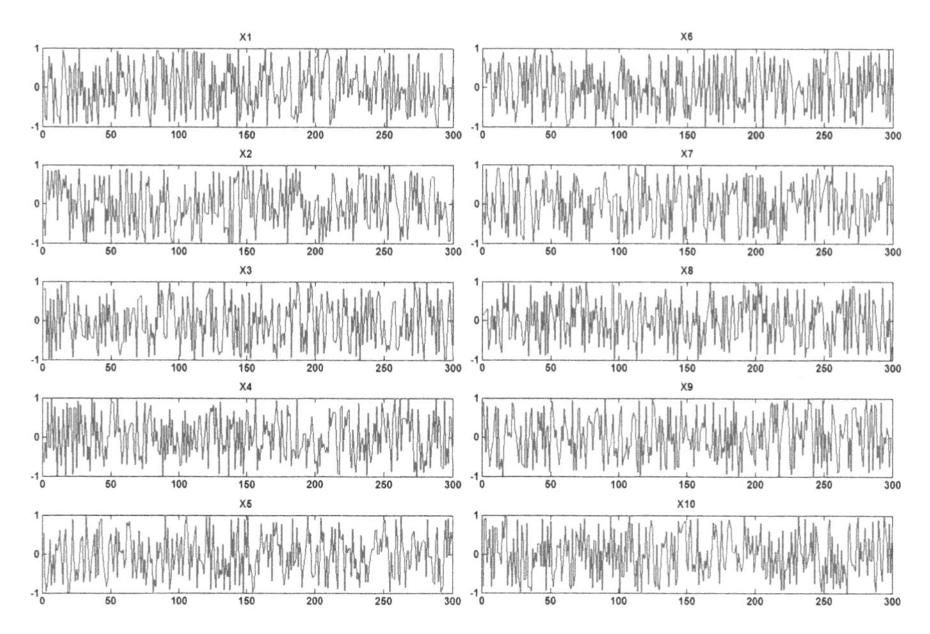

Fig. 3.2 Illustration of linear regression: input test data

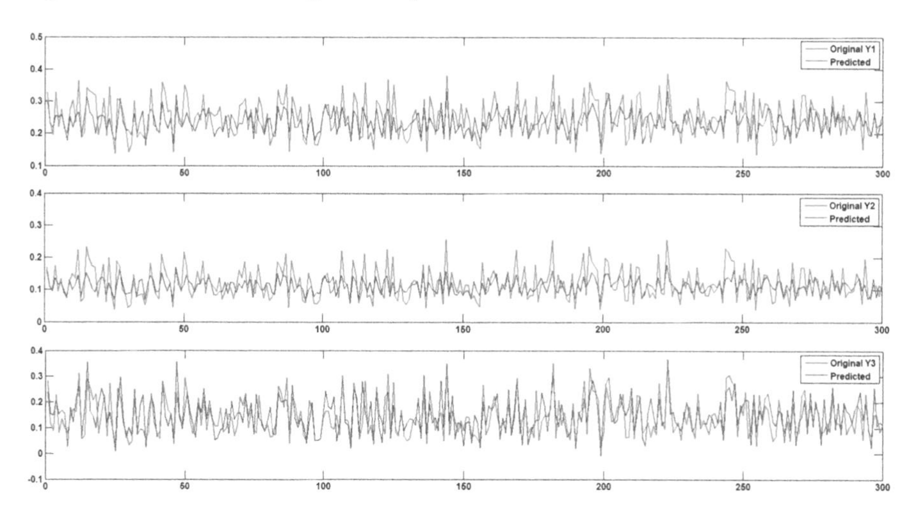

Fig. 3.3 Illustration of linear regression: output test data and the corresponding predicted data

is treated as the constant 1. The other basis functions ($k_i, i = 1 \cdots 3$) are chosen as Gaussian basis (with three distinct mean vectors $m_i, i = 1 \cdots 3$).

$$k_i(\mathbf{x}, \mathbf{m_i}) = e^{\frac{(\mathbf{x}-\mathbf{m_i})^T(\mathbf{x}-\mathbf{m_i})}{2\sigma^2}} \tag{3.22}$$

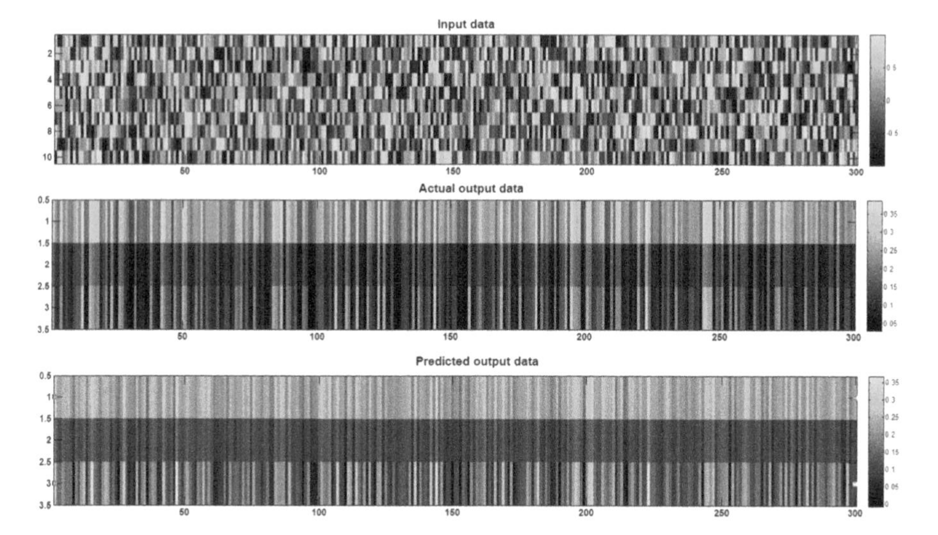

Fig. 3.4 Illustration of linear regression: input test data, output test data, and the corresponding predicted data

regressiondemo.m

```
%Data generation with noise
W1=rand(10,5);
B1=rand(5,1);
W2=rand(5,3);
B2=rand(3,1);
INPUT=rand(1000,10)*2-1;
H=INPUT*W1+repmat(B1',1000,1);
H=1./(1+exp(H))
OUTPUT=H*W2+repmat(B2',1000,1);
OUTPUT=1./(1+exp(OUTPUT));
INPUTTRAIN=INPUT(1:1:700,:);
INPUTTEST=INPUT(701:1:1000,:);
OUTPUTTRAIN=OUTPUT(1:1:700,:)+randn(700,3)*0.2;
OUTPUTTEST=OUTPUT(701:1:1000,:);

figure(1)
colormap(copper)
subplot(3,1,1)
imagesc(INPUTTRAIN');
subplot(3,1,2)
imagesc(OUTPUT(1:1:700,:)')
subplot(3,1,3)
imagesc(OUTPUTTRAIN')
```

```
%Using Gaussian kernel based polynomial regression
[V,M]=kmeans(INPUTTRAIN,4);
[V1,I1]=min(sum((INPUTTRAIN-repmat(M(1,:),
700,1)).^2));
[V2,I2]=min(sum((INPUTTRAIN-repmat(M(2,:),
700,1)).^2));
[V3,I3]=min(sum((INPUTTRAIN-repmat(M(3,:),
700,1)).^2));
[V4,I4]=min(sum((INPUTTRAIN-repmat(M(3,:),
700,1)).^2));
sigma=mean([V1 V2 V3 V4]);
B1=[];
B2=[];
B3=[];
B4=[];
for i=1:1:size(INPUTTRAIN,1)
B1=[B1 GBF(INPUTTRAIN(i,:),M(1,:),sigma)];
B2=[B2 GBF(INPUTTRAIN(i,:),M(2,:),sigma)];
B3=[B3 GBF(INPUTTRAIN(i,:),M(3,:),sigma)];
B4=[B4 GBF(INPUTTRAIN(i,:),M(4,:),sigma)];
end
%Formulation of matrix
MAT=[ones(700,1) B1' B2' B3' B4'];
W=pinv(MAT)*OUTPUTTRAIN;
save W W
%TESTING
B1=[];
B2=[];
B3=[];
B4=[];
for i=1:1:size(INPUTTEST,1)
B1=[B1 GBF(INPUTTEST(i,:),M(1,:),sigma)];
B2=[B2 GBF(INPUTTEST(i,:),M(2,:),sigma)];
B3=[B3 GBF(INPUTTEST(i,:),M(3,:),sigma)];
B4=[B4 GBF(INPUTTEST(i,:),M(4,:),sigma)];
end
MAT=[ones(300,1) B1' B2' B3' B4']
OUTPUTTESTR=MAT*W;

figure(2)
subplot(5,2,1)
plot(INPUTTEST(:,1))
subplot(5,2,2)
plot(INPUTTEST(:,2))
subplot(5,2,3)
```

```
plot(INPUTTEST(:,3))
subplot(5,2,4)
plot(INPUTTEST(:,4))
subplot(5,2,5)
plot(INPUTTEST(:,5))
subplot(5,2,6)
plot(INPUTTEST(:,6))
subplot(5,2,7)
plot(INPUTTEST(:,7))
subplot(5,2,8)
plot(INPUTTEST(:,8))
subplot(5,2,9)
plot(INPUTTEST(:,9))
subplot(5,2,10)
plot(INPUTTEST(:,10))


figure(3)
subplot(3,1,1)
plot(OUTPUTTEST(:,1))
hold on
plot(OUTPUTTESTR(:,1))
subplot(3,1,2)
plot(OUTPUTTEST(:,2))
hold on
plot(OUTPUTTESTR(:,2))
subplot(3,1,3)
plot(OUTPUTTEST(:,3))
hold on
plot(OUTPUTTESTR(:,3))

figure(4)
colormap(copper)
subplot(3,1,1)
imagesc(INPUTTEST')
subplot(3,1,2)
imagesc(OUTPUTTEST')
subplot(3,1,3)
imagesc(OUTPUTTESTR')

%Prior probability density function of INPUT data is
%uniform distribution between -1 to 1
%Likelihood function of output given INPUT data
%Given the first INPUT data, output is distributed as
follows.
```

```
figure(5)
colormap(copper)
OUTPUTNOISE=[];
for i=1:1:1000
    OUTPUTNOISE=[OUTPUTNOISE; OUTPUT(1,:)+randn(1,3)*
    0.2];
end
scatter3(OUTPUTNOISE(:,1),OUTPUTNOISE(:,2),
OUTPUTNOISE(:,3),'g*');
hold on
M1=mean(OUTPUTNOISE);
scatter3(M1(1),M1(2),M1(3),'or');
hold on
scatter3(OUTPUT(1,1),OUTPUT(1,2),OUTPUT(1,3),'+r');

figure(6)
colormap(copper)
OUTPUTNOISE=[];
for i=1:1:1000
    OUTPUTNOISE=[OUTPUTNOISE; OUTPUT(2,:)+randn(1,3)*
    0.2];
end
scatter3(OUTPUTNOISE(:,1),OUTPUTNOISE(:,2),
OUTPUTNOISE(:,3),'g*');
M2=mean(OUTPUTNOISE);
hold on
scatter3(M2(1),M2(2),M2(3),'or');
hold on
scatter3(OUTPUT(2,1),OUTPUT(2,2),OUTPUT(2,3),'+r');

figure(7)
colormap(copper)
OUTPUTNOISE=[];
for i=1:1:1000
    OUTPUTNOISE=[OUTPUTNOISE; OUTPUT(3,:)+randn(1,3)*
    0.2];
end

scatter3(OUTPUTNOISE(:,1),OUTPUTNOISE(:,2),
OUTPUTNOISE(:,3),'g*');
M3=mean(OUTPUTNOISE);
hold on
scatter3(M3(1),M3(2),M3(3),'or');
hold on
scatter3(OUTPUT(3,1),OUTPUT(3,2),OUTPUT(3,3),'+r');
```

3.2.2 Least Square Technique Versus Maximum Likelihood Estimation

Let rows of the matrix $\mathbf{E}$ are Gaussian distributed with mean zero vector and co-variance matrix $\mathbf{I}\frac{1}{\beta}$. Hence i^{th} row of the matrix $\mathbf{T}$ in $\mathbf{T} = \mathbf{F}\mathbf{W}^T + \mathbf{E}$ is Gaussian distributed with mean vector $\mathbf{f_i}\mathbf{W^T}$ and co-variance matrix $\mathbf{I}\frac{1}{\beta}$, where $\mathbf{f_i}$ is the ith row of the matrix $\mathbf{F}$. The joint density function of the random vectors obtained from the rows of the matrix $\mathbf{T}$ is the product of the individual joint density function of each row and is given as follows:

$$K\Pi_{i=1}^{i=d} e^{-|\mathbf{t_i}-\mathbf{f_i}\mathbf{W^T}|^2} \tag{3.23}$$

where K is the constant.

This is the likelihood function of the $\mathbf{T}$ given $\mathbf{W}$ (with fixed input training data). We need to estimate $\mathbf{W}$ by maximizing the likelihood function. As the likelihood function is the increasing function, we can estimate $\mathbf{W}$ by maximizing the logarithm of the likelihood function as follows:

$$\sum_{i=1}^{i=d} -|\mathbf{t_i} - \mathbf{f_i}\mathbf{W^T}|^2 \tag{3.24}$$

Equivalently, we estimate $\mathbf{W}$ by minimizing the function as follows:

$$\sum_{i=1}^{i=d} |\mathbf{t_i} - \mathbf{f_i}\mathbf{W^T}|^2 = ||\mathbf{T} - \mathbf{F}\mathbf{W}^T||_2^2. \tag{3.25}$$

Thus the estimation obtained by using least square technique and by maximizing the likelihood function ends up with an identical solution.

3.3 Inference on Figs. 3.5, 3.6, and 3.7

The random variable $\mathbf{t}$ is the noisy observation and is modeled as $\mathbf{y}(\mathbf{x}) + \boldsymbol{\epsilon}$, where $\boldsymbol{\epsilon}$ is the additive Gaussian noise. Let the scatter plots (1000 green points) in Figs. 3.5, 3.6, and 3.7 be the noisy observations corresponding to the single point $\mathbf{y}(\mathbf{x_1})$, $\mathbf{y}(\mathbf{x_2})$, and $\mathbf{y}(\mathbf{x_3})$, respectively. The estimate of mean of the points in the scatter plot given the particular input $\mathbf{x_i}$, i.e., the conditional mean of the target vector $\mathbf{t_i}$ given $\mathbf{x_i}$, is represented as o and the actual output $\mathbf{y}(\mathbf{x_i})$ is represented as $+$. It is seen from the figure that the points o and $+$ are closer to each other. Thus it is observed that the conditional mean of the random vector $\mathbf{t_i}$ given $\mathbf{x_i}$ is the estimate of the true value of the output.

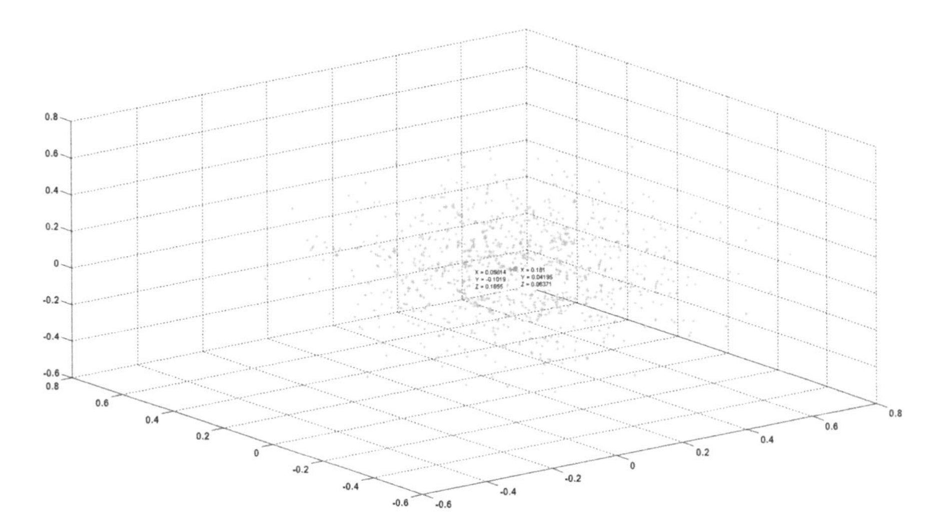

Fig. 3.5 Illustrates that the parametric least square solution is closer to the result obtained as the conditional mean of the likelihood function (data 1). *Note*: Predicted point using linear regression [0.1649 0.0732 0.0113] is closer to the estimated mean of the data. The result of the regression technique is closer to the result obtained as the conditional mean of the likelihood function $f(t/x)$

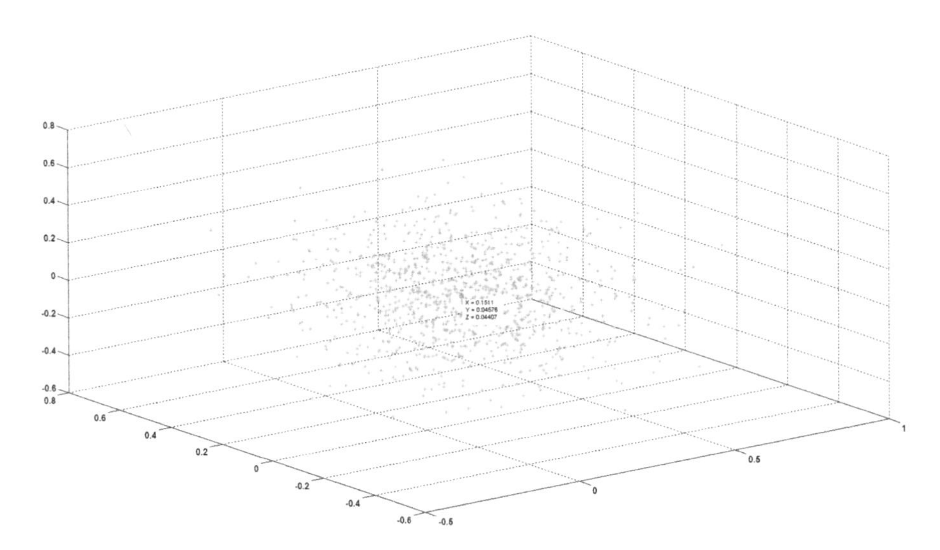

Fig. 3.6 Illustrates that the parametric linear regression solution is closer to the result obtained as the conditional mean of the likelihood function (data 2). *Note*: Predicted point using linear regression [0.2044 0.0928 0.0926] is closer to the estimated mean of the data. The result of the regression technique is closer to the result obtained as the conditional mean of the likelihood function $f(t/x)$

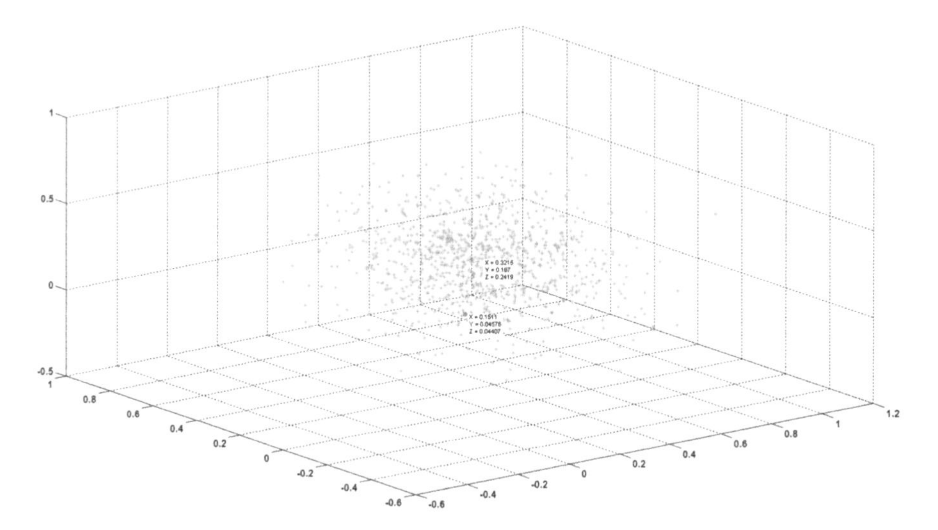

Fig. 3.7 Illustrates that the linear regression is closer to the result obtained as the conditional mean of the likelihood function (data 3). *Note*: Predicted point using linear regression [0.2563 0.1390 0.2614] is closer to the estimated mean of the data. The result of the regression technique is closer to the result obtained as the conditional mean of the likelihood function $f(t/x)$

3.4 Regularization

The matrix $\mathbf{W}$ is obtained by minimizing $||\mathbf{T} - \mathbf{F}\mathbf{W}^T||_2$. The matrix $\mathbf{F}$ and $\mathbf{T}$ are constructed using the training data set, i.e., for the specific set of input vectors $(\mathbf{x_1}, \mathbf{x_2}, ..., \mathbf{x_d})$ and the corresponding target vectors $(\mathbf{t_1}, \mathbf{t_2}, ..., \mathbf{t_d})$. If the training is done such that the mean square error is almost zero for the training data, the error obtained using the testing data gets deviated largely. This is the situation of over-fitting the training model. To circumvent this case, the regularization factor is added in the objective function as follows: $||\mathbf{T} - \mathbf{F}\mathbf{W}^T||_2^2 + \lambda||\mathbf{W}||_2^2$, where λ is the regularization constant.

The solution to $\mathbf{W}$ is obtained as follows:

$$\widehat{\mathbf{W}}^T = (\mathbf{F}^T\mathbf{F} + \lambda\mathbf{I})^{-1}\mathbf{F}^T\mathbf{T} \tag{3.26}$$

3.4.1 Illustration on Tuning Regularization Constant

The experiment is performed with single-input, single-output case. Let the number of data points collected be 8000. Divide them into two groups: (a) training data with 4000 data points and (b) testing data with 4000 data points.

1. Divide the training data points into 40 groups with 100 data points in each group. Let it be $\mathbf{g_1}, \mathbf{g_2}, \ldots, \mathbf{g_{40}}$.
2. Fix the regularization constant as λ.
3. The estimated $\mathbf{test_o}$ (using the model obtained with i^{th} training set g_i) is represented as $\mathbf{est_i}$.
4. Mean estimate is obtained as follows:

$$\mathbf{meanest} = \frac{1}{7}\sum_{i=1}^{i=7} \mathbf{est_i} \tag{3.27}$$

5. The metric $Bias^2$ is used to check how the mean vector **meanest** is deviated from the actual output vector $\mathbf{test_o}$ and is computed as follows:

$$Bias^2 = \frac{1}{100}\sum_{n=1}^{n=100} (test_o(n) - meanest(n))^2 \tag{3.28}$$

6. The metric var is used to check how the estimated vectors ($\mathbf{est_i}$), for $i = 1 \cdots N$, are closer to each other and is computed as follows:

$$var = \frac{1}{(40 \times 100)}\sum_{i=1}^{i=40} (\mathbf{est_i} - \mathbf{meanest})^T(\mathbf{est_i} - \mathbf{meanest}) \tag{3.29}$$

7. Repeat steps 3–6 by fixing various values of λ. Choose the λ corresponding to the lowest value of $Bias^2 + var$ (refer Fig. 3.8). It is observed from Fig. 3.9 that $Bias^2$ is directly proportional to λ, but var is inversely proportional to λ.

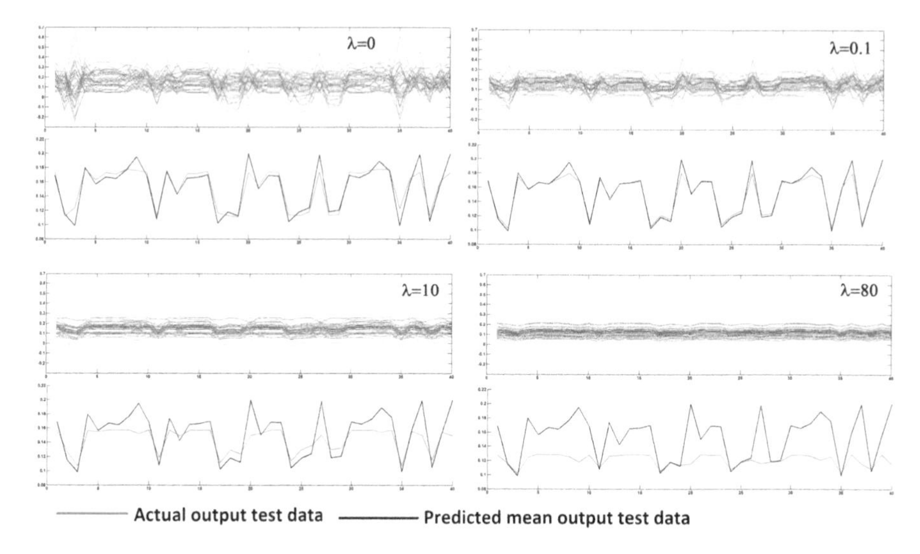

Fig. 3.8 Illustration of 40 estimates of output test data with various regularization constant. The mean estimate (black) and the corresponding actual test output (red) are given for illustration

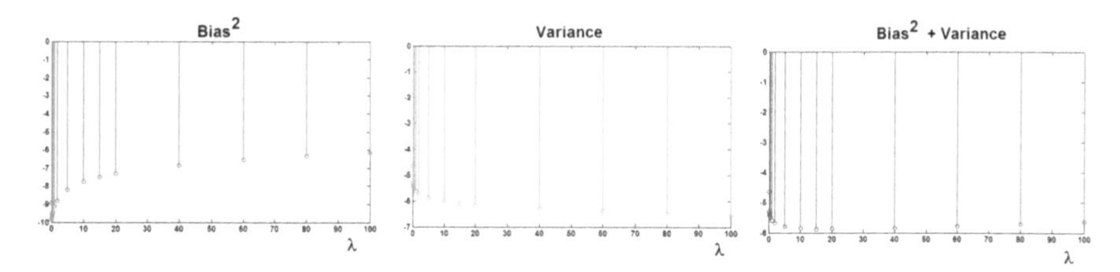

Fig. 3.9 Illustration of the effect of regularization constant on $Bias^2$ and var using 40 training sets. It is seen that $Bias^2$ is increasing with increase in regularization constant and the var is decreasing with increase in regularization constant

3.5 Importance of $Bias^2 + var$

Let t be the random variable associated with observation given the input random variable x. The estimate of the random variable t given x is estimated as conditional mean of the posterior density function $f(t/x)$, i.e., $g(x) = E(t/x)$. It is understood in this case that the prior density function is associated with t, i.e., $f(t)$, and the posterior density function is represented as $f(t/x)$. Let $y(x)$ is the estimate obtained using linear regression. Thus the error involved in estimating the actual output using regression model $(y(x))$ is given as follows:

$$J = E((y(x) - t)^2) \tag{3.30}$$

The error involved in estimation is represented as follows:

$$E((y(x) - g(x) + g(x) - t)^2) \tag{3.31}$$

$$= E((y(x) - g(x))^2) + E((g(x) - t)^2) + E(2(y(x) - g(x))(g(x) - t)) \tag{3.32}$$

It is observed that the expectation is performed using the joint density function $f(x, t)$. Consider the third term as follows:

$$E(2(y(x) - g(x))(g(x) - t)) \tag{3.33}$$

$$= E_x(E(2(y(x) - g(x))(g(x) - t))/x) \tag{3.34}$$

$$= E_x(2(y(x) - g(x))E((g(x) - t)/x)) \tag{3.35}$$

$$= 0 \tag{3.36}$$

as $E((g(x) - t)/x) = E(g(x)/x) - E(t/x) = g(x) - g(x) = 0$. The second term $E((t - g(x))^2)$ is the error involved due to additive noise. This cannot be reduced. This is the contribution of the regression error due to the additive noise.

Consider the first term: $E((y(x) - g(x))^2)$. Let $y_D(x)$ be the regression curve obtained using the typical data set D.

$$\begin{aligned}
&E((y(x) - g(x))^2) \\
&= E_x E_D((y_D(x) - g(x))^2) \\
&= E_x E_D((y_D(x) - E_D((y_D(x))) + E_D((y_D(x))) - g(x))^2) \\
&= E_x E_D((y_D(x) - E_D((y_D(x))))^2) \\
&\quad + E_x((E_D((y_D(x))) - g(x))^2) \\
&\quad + 2E_x E_D((y_D(x) - E_D((y_D(x))))(y_D(x) - g(x)))
\end{aligned}$$

It is noted that

$$E_x E_D((y_D(x) - E_D((y_D(x)))(y_D(x) - g(x))) = 0 \tag{3.37}$$

Also the first term

$$E_x E_D((y_D(x) - E_D((y_D(x))))^2) \tag{3.38}$$

is the variance represented as var and the second term

$$E_x((E_D((y_D(x))) - g(x))^2) \tag{3.39}$$

is $Bias^2$. Thus the error involved in linear regression is represented as the summation of the var, $Bias^2$, and noise. Thus the optimal hyperparameter (e.g., λ in regularization) is obtained when $var + Bias^2$ is minimized.

Biasvardemo.m

```
%Bias variance Normalization demo
W1=rand(1,5);
B1=rand(5,1);
W2=rand(5,1);
B2=rand(1,1);

%DATA1
INPUT=rand(4000,1)*2-1;
H=INPUT*W1+repmat(B1',4000,1);
H=1./(1+exp(H));
OUTPUT=H*W2+repmat(B2',4000,1);
OUTPUT=1./(1+exp(OUTPUT));
for i=1:1:40
INPUTTRAIN{i}=INPUT((i-1)*100+1:1:(i-1)*100+100,:);
OUTPUTTRAIN{i}=OUTPUT((i-1)*100+1:1:(i-1)*100+100,:)+
randn(100,1)*sqrt(0.2);
end

INPUTTEST=rand(40,1)*2-1;
H=INPUTTEST*W1+repmat(B1',40,1);
H=1./(1+exp(H));
OUTPUT=H*W2+repmat(B2',40,1);
OUTPUTTEST=1./(1+exp(OUTPUT));

sigma=1.2;
SSECOL=[];
VAR=[];
BIAS=[];
k=0;
REG=[0 0.1:0.1:0.5 1 2 5 10 15 20:20:100];
for i=1:1:length(REG)
    k=k+1;
    for j=1:1:40
[temp,MSE,W]=performregression(INPUTTRAIN{j},
OUTPUTTRAIN{j},...
INPUTTEST,OUTPUTTEST,REG(i),sigma);
OUTPUTVALR{j}=temp';
figure(k)
subplot(2,1,1)
plot(OUTPUTVALR{j},'b')
```

```
hold on
    end
subplot(2,1,2)
MEANESTIMATE=mean(cell2mat(OUTPUTVALR'));
hold on
plot(MEANESTIMATE,'r')
hold on
plot(OUTPUTTEST,'k')
BIAS=[BIAS sum((OUTPUTTEST'-MEANESTIMATE).^2)/length
(OUTPUTTEST)];
VAR=[VAR  mean(var(cell2mat(OUTPUTVALR')))];
end
figure
subplot(2,2,1)
stem(REG,log(BIAS),'r')
subplot(2,2,2)
stem(REG,log(VAR),'g')
subplot(2,2,3)
stem(REG,log(BIAS+VAR),'b')
[P1,Q1]=min(BIAS+VAR);
MSECOL=[];
for i=1:1:length(REG)
[OUTPUTTESTR,MSE,W]=performregression(INPUTTRAIN{1},
OUTPUTTRAIN{1},...
INPUTTEST,OUTPUTTEST,REG(i),sigma);
MSECOL=[MSECOL MSE];
end
subplot(2,2,4)
stem(REG,log(MSECOL),'k')
```

performregression.m

```
function [OUTPUTTESTR,MSE,W1]=performregression
(INPUTTRAIN,OUTPUTTRAIN,...
INPUTTEST,OUTPUTTEST,reg,sigma)
[V,M]=kmeans(INPUTTRAIN,4);
[V1,I1]=min(sum((INPUTTRAIN-repmat(M(1,:),size
(INPUTTRAIN,1),1)).^2));
[V2,I2]=min(sum((INPUTTRAIN-repmat(M(2,:),size
(INPUTTRAIN,1),1)).^2));
[V3,I3]=min(sum((INPUTTRAIN-repmat(M(3,:),size
(INPUTTRAIN,1),1)).^2));
[V4,I4]=min(sum((INPUTTRAIN-repmat(M(4,:),size
(INPUTTRAIN,1),1)).^2));
B1=[];
B2=[];
```

```
B3=[];
B4=[];
for i=1:1:size(INPUTTRAIN,1)
B1=[B1 gbf(INPUTTRAIN(i,:),M(1,:),sigma)];
B2=[B2 gbf(INPUTTRAIN(i,:),M(2,:),sigma)];
B3=[B3 gbf(INPUTTRAIN(i,:),M(3,:),sigma)];
B4=[B4 gbf(INPUTTRAIN(i,:),M(4,:),sigma)];
end
%Formulation of matrix
MAT=[ones(size(INPUTTRAIN,1),1) B1' B2' B3' B4'];
MAT1=MAT'*MAT;
W1=pinv(MAT1+reg*eye(size(MAT1,1)))*MAT'*OUTPUTTRAIN;
B1=[];
B2=[];
B3=[];
B4=[];
for i=1:1:size(INPUTTEST,1)
B1=[B1 gbf(INPUTTEST(i,:),M(3,:),sigma)];
B2=[B2 gbf(INPUTTEST(i,:),M(2,:),sigma)];
B3=[B3 gbf(INPUTTEST(i,:),M(3,:),sigma)];
B4=[B4 gbf(INPUTTEST(i,:),M(4,:),sigma)];
end
MAT=[ones(size(INPUTTEST,1),1) B1' B2' B3' B4'];
OUTPUTTESTR=MAT*W1;
MSE=sum((OUTPUTTESTR-OUTPUTTEST).^2)/length
(OUTPUTTEST);
```

3.6 Bayesian Approach

In the parametric approach, the least square solution is obtained by solving $\mathbf{T} = \mathbf{FW}$, where $\mathbf{T}$ is the target matrix with size $d \times n$, where d is the number of training data and n is the size of the target vector corresponding to the input vector $\mathbf{x}$. The least square solution to estimate is obtained as $\mathbf{W_{cap}} = (\mathbf{F^T F})^{-1}(\mathbf{F^T})\mathbf{T}$. Let $\mathbf{T}$ is the noisy observation matrix of the output (i.e., actual output with additive multivariate Gaussian distributed noise). The likelihood function $p(\mathbf{T}/\mathbf{W})$ to estimate $\mathbf{W}$ is modeled as multivariate Gaussian distributed, and the maximum likelihood estimation is given as $\mathbf{W_{cap}} = (\mathbf{F^T F})^{-1}(\mathbf{F^T})\mathbf{T}$. The columns of the matrix $\mathbf{W}$ are Gaussian distributed with mean vector 0 and co-variance matrix $\frac{1}{\alpha}$. Hence the posterior density function $p(\mathbf{W}/\mathbf{T})$ is given as the multivariate Gaussian distributed with mean of the matrix $\mathbf{M} = (\alpha I + \beta \mathbf{F}^T \mathbf{F})^{-1} \beta \mathbf{F}^T \mathbf{T} = (\mathbf{F}^T \mathbf{F} + \frac{\alpha}{\beta} I)^{-1} (\mathbf{F}^T)\mathbf{T}$. If the a posteriori density function is Gaussian distributed, the conditional mean is obtained as the mean of the posterior density function. Thus the estimate of the weight matrix is obtained as $(\mathbf{F}^T\mathbf{F} + \frac{\alpha}{\beta}I)^{-1}(\mathbf{F}^T)\mathbf{T}$. Hence Bayes solution is considered as the special case of the solution obtained using the regularization technique with regularization constant ($\lambda = \frac{\alpha}{\beta}$).

3.7 Observations on Figs. 3.10, 3.11, 3.12, and 3.13

The illustration is done with $m = 1$ and $n = 1$. The number of elements of the vector $\mathbf{w}$ is two. The prior density function of the random vector is Gaussian distributed. The contour plot of the typical outcome of the random vector $\mathbf{w}$ with $\alpha = 0.1$ is plotted in Fig. 3.10. The number of outcomes is considered as 1000. The normalized histogram of the outcomes is used to fill matrix of size 50×50. Given the first data x_1, the outcome of the likelihood function $p(t/\mathbf{w}, x)$ is obtained as $\mathbf{w}^T f(x_1) + \epsilon$, where $\mathbf{w}$ is obtained as Gaussian distributed with mean zero and variance $\frac{1}{\alpha}$. Similarly, the ϵ (noise) is modeled as Gaussian distributed with mean zero and variance $\frac{1}{\beta}$. The normalized histogram of $\mathbf{t}$ as the function of $\mathbf{w}$ is filled in the matrix of size 50×50 to obtain the likelihood matrix (refer Fig. 3.11). The prior density matrix and the likelihood matrix (both as the function of $\mathbf{w}$) are multiplied to obtain the a posteriori density function as the matrix (refer Fig. 3.12). This is repeated for the 10 data and the value of $\mathbf{w}$ is converging to the single point (refer Fig. 3.13). The point thus obtained is the solution for $\mathbf{w}$ using Bayes technique.

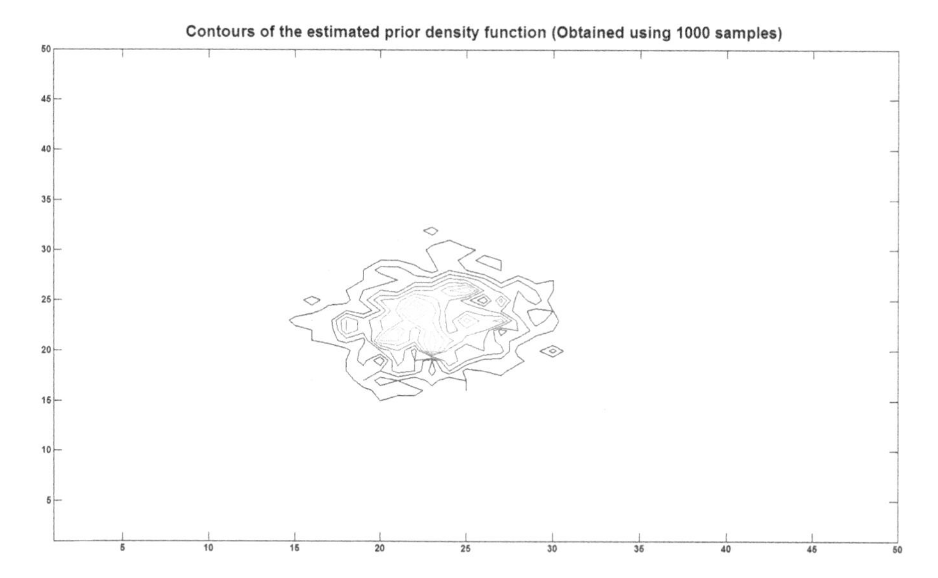

Fig. 3.10 Prior distribution of the random vector w

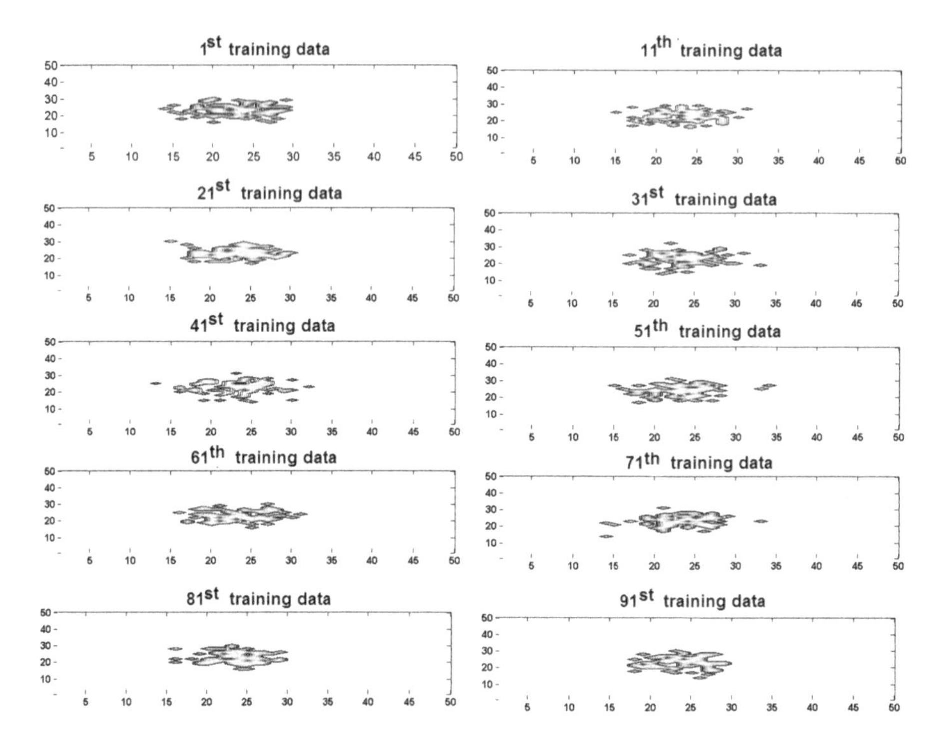

Fig. 3.11 Likelihood function of w given input vector

Bayesianapproachdemo.m

```
%Bayesian approach
% clc
alpha=0.1;
beta=0.2;
W1=rand(1,2);
B1=rand(2,1);
W2=rand(2,1);
B2=rand(1,1);
save WEIGTHS W1 B1 W2 B2
%DATA GENERATION
INPUT=rand(1000,1)*2-1;
H=INPUT*W1+repmat(B1',1000,1);
H=1./(1+exp(H));
OUTPUT=H*W2+repmat(B2',1000,1);
OUTPUT=1./(1+exp(OUTPUT))+randn(1000,1)*beta;
M=mean(INPUT');
sigma=var(INPUT');
B1=[];
```

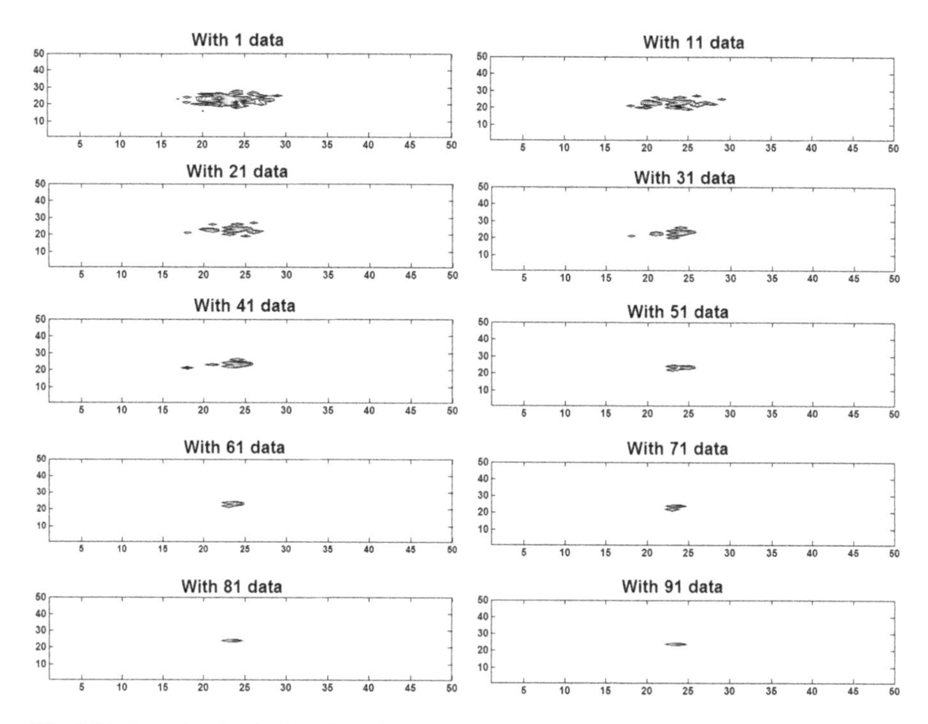

Fig. 3.12 Posterior density function of the random vector w given input training data

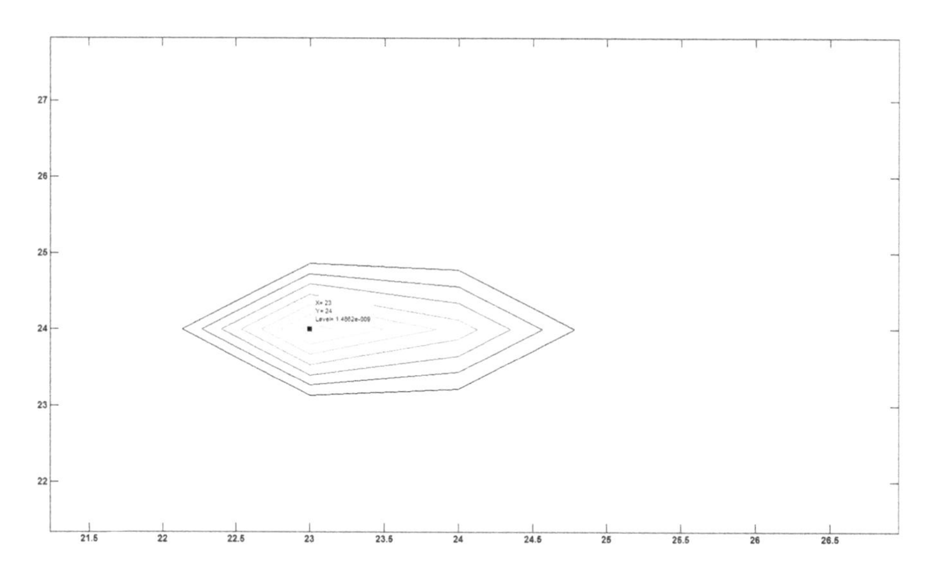

Fig. 3.13 Final convergence of the random vector w

```
for i=1:1:size(INPUT,1)
B1=[B1 gbf(INPUT(i,:),M',sigma)];
end
%Formulation of matrix
MAT=[ones(size(INPUT,1),1) B1'];
MAT1=MAT'*MAT;
save MAT MAT
W=pinv(MAT1+(alpha/beta)*eye(size(MAT1,1)))*MAT'*
OUTPUT;
save W W
%%%%%%%%%%%%%%%%%%%%%%%%%%%%%%%%%%%%%%%%%%%%%%%%%%%%
%Estimating W using Bayesian approach
%For the first input (First row of MAT file)
load W
load MAT
WDATA=randn(2,10000)*sqrt(alpha);
%ma=max(max(WDATA));
mi=min(min(WDATA));
mibias=abs(round((mi*10)));
%mabias=abs(round((ma*10)));
PRIOR=zeros(50,50);
for i=1:1:1000
PRIOR(round(WDATA(1,i)*10)+mibias+10,round(WDATA(2,i)
*10)+mibias+10)...
=PRIOR(round(WDATA(1,i)*10)+mibias+10,round(WDATA(2,i)
*10)+mibias+10)+1;
end
PRIOR=PRIOR/1000;
POST=PRIOR;
figure(1)
contour(PRIOR)
LFMAT=zeros(50,50);
COL=[];
sub=0;

for i=1:1:10
sub=sub+1;
for p=1:1:10
LFMAT=zeros(size(POST,1),size(POST,2));
for j=1:1:100
W1=randn(2,1)*sqrt(alpha);
COL=[];
for k=1:1:10000
COL=[COL MAT(i,:)*W1+randn(1)*beta];
end
```

```
h=hist(COL,15);
temp=h/sum(h);
[U,V]=max(temp);
LFMAT(round(W1(1)*10)+mibias+10,round(W1(2)*10)+mibias
+10)=temp(V);
end
end
figure(2)
subplot(5,2,sub)
contour(LFMAT)
pause(0.1)
POST=LFMAT.*POST
figure(3)
subplot(5,2,sub)
contour(POST)
pause(0.1)
end
figure(4)
contour(POST)
```

gbf.m

```
function [res]=gbf(i,m,s)
res=exp(-((i-m)*(i-m)')/(2*s^2));
```

gausskernelmethod.m

```
%Regression using gaussian kernel method
W1=rand(3,2);
B1=rand(2,1);
W2=rand(2,1);
B2=rand(1,1);
%DATA
N=1000;
INPUT=randn(N,3)*2-1;
H=INPUT*W1+repmat(B1',N,1);
H=1./(1+exp(H));
OUTPUT=H*W2+repmat(B2',N,1);
OUTPUT=1./(1+exp(OUTPUT));
INPUTTRAIN=INPUT(1:1:400,:);
INPUTVAL=INPUT(401:1:700,:);
INPUTTEST=INPUT(701:1:1000,:);
OUTPUTTRAIN=OUTPUT(1:1:400,:)+randn(400,1)*0.02;
OUTPUTVAL=OUTPUT(401:1:700,:);
OUTPUTTEST=OUTPUT(701:1:1000,:);
%Formulation of smoother matrix
s=[0.1:0.1:5];
```

```
OUTPUTest=[];
MSE=[];
temp=0;
KERNELVAL=[];
for k=1:1:length(s)
    OUTPUTVALest=[];
    for i=1:1:size(INPUTVAL,1)
        temp=[];
        KERNELVAL=[];
     for j=1:1:size(INPUTTRAIN,1)
         KERNELVAL=[KERNELVAL gbf(INPUTTRAIN(j,:),
         INPUTVAL(i,:),s(k))];
         KERNELVAL=KERNELVAL/sum(KERNELVAL);
     end
    temp=sum(KERNELVAL*OUTPUTTRAIN(:,1));
    OUTPUTVALest=[OUTPUTVALest temp];
    end
  MSE=[MSE sum(sum((OUTPUTVAL-OUTPUTVALest).^2))/
  length(OUTPUTVAL)];
end
[P,Q]=min(MSE);
ssel=s(Q);
OUTPUTTESTest=[];
MSETEST=[];
for i=1:1:size(INPUTTEST,1)
        temp=[];;
        KERNELTEST=[];
   for j=1:1:size(INPUTTRAIN,1)
        KERNELTEST=[KERNELTEST gbf(INPUTTRAIN(j,:),
        INPUTTEST(i,:),ssel)];
        KERNELTEST=KERNELTEST/sum(KERNELTEST);
   end
    temp=sum(KERNELTEST*OUTPUTTRAIN(:,1));
    OUTPUTTESTest=[OUTPUTTESTest temp];
end
MSETEST=sum(sum((OUTPUTTEST'-OUTPUTTESTest).^2))/
length(OUTPUTTEST);
figure
subplot(2,1,1)
plot(OUTPUTTEST,'b')
subplot(2,1,2)
plot(OUTPUTTESTest,'r')
```

3.8 Linear Regression Using Iterative Method

The solution that minimizes $\mathbf{J} = ||\mathbf{T} - \mathbf{FW}||^2$ is obtained as $(\mathbf{F^T F})^{-1}\mathbf{F^T T}$. As the size of the matrix $\mathbf{F^T F}$ is large, the complexity in computing the inverse increases. The optimal $\mathbf{W}$ that minimizes $\mathbf{J}$ is obtained using gradient-descent algorithm. Suppose we would like to find out the values for x that minimizes the function $f(x)$. The Taylor series expansion for the function $f(x)$ is given as follows:

$$f(x+h) = f(x) + \frac{h}{1!}f'(x) + \frac{h^2}{2!}f''(x) + \cdots \tag{3.40}$$

Suppose we choose $h = -kf'(x)$, we get the following:

$$f(x - kf'(x)) = f(x) - k\frac{|f'(x)|^2}{1!} + \cdots \tag{3.41}$$

$$\Rightarrow f(x - kf'(x)) \leq f(x) \tag{3.42}$$

Thus if we start with some initial guess on the value x, the next choice for x that minimizes the function $f(x)$ is given as follows:

$$x_{next} = x_{current} - kf'(x_{current}) \tag{3.43}$$

where k is the constant. This technique is used to obtain the optimal value of $\mathbf{W}$ that minimizes $\mathbf{J} = ||\mathbf{T} - \mathbf{FW}||^2$ as follows:

$$\mathbf{W}(iter+1) = \mathbf{W}(iter) - \eta\mathbf{\Delta J} \Rightarrow \mathbf{W}(iter+1) = \mathbf{W}(iter) + \eta\mathbf{F}^T(\mathbf{T} - \mathbf{FW}) \tag{3.44}$$

where $iter$ is the iteration number.

In practice, the update of W is done for the individual training set one after another as follows:

$$\mathbf{W}(t+1) = \mathbf{W}(t) + \eta f(\mathbf{x_k})(\mathbf{t_k} - \mathbf{f(x_k)^T W}) \tag{3.45}$$

where $\mathbf{t_k}$ is the kth row of the matrix T with size $1 \times n$. The size of the vector $f(x_k)$ is $M \times 1$. The steps involved in sequential learning are summarized as follows:

1. Let $\mathbf{x_1}, \mathbf{x_2}, \ldots, \mathbf{x_d}$ be the training vectors.
2. Initialize the weight matrix $\mathbf{W}(1)$ and initialize the value for $iter = 1$ and $k = 1$.
3. Update the weight matrix $\mathbf{W}(iter+1) = \mathbf{W}(iter) + \eta f(\mathbf{x_k})(\mathbf{t_k} - \mathbf{f(x_k)^T W})$.
4. Update $iter = iter + 1$ and $k = k + 1$.
5. Repeat (3) until $iter = d$. If $iter = d$, reset $iter = 1$ and $k = 1$. This is one epoch.
6. Compute the Sum Squared Error $||\mathbf{T} - \mathbf{FW}||^2$, using the latest updated matrix $\mathbf{W}$ after one epoch.

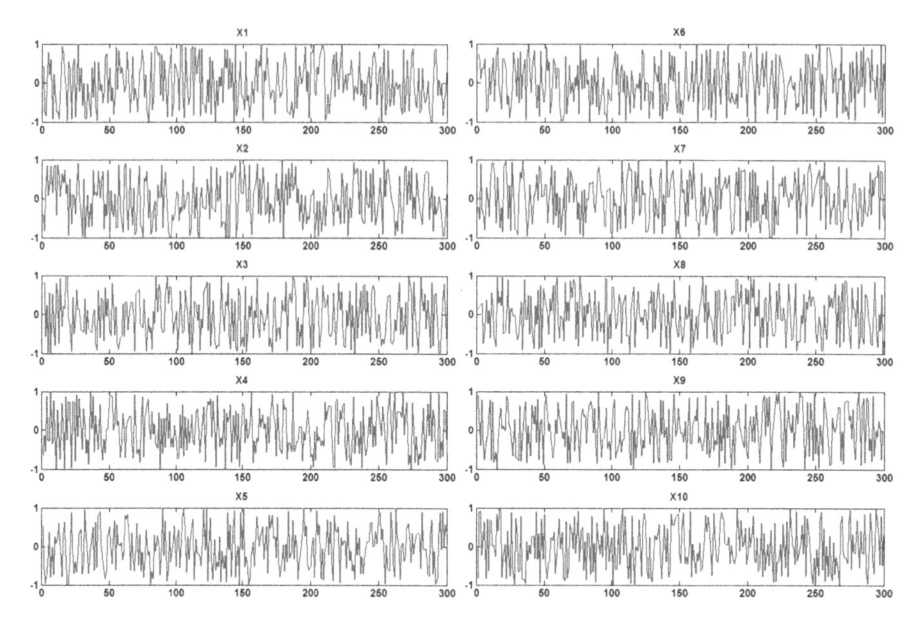

Fig. 3.14 Testing data (input) used to demonstrate iterative (steepest-descent) based regression technique

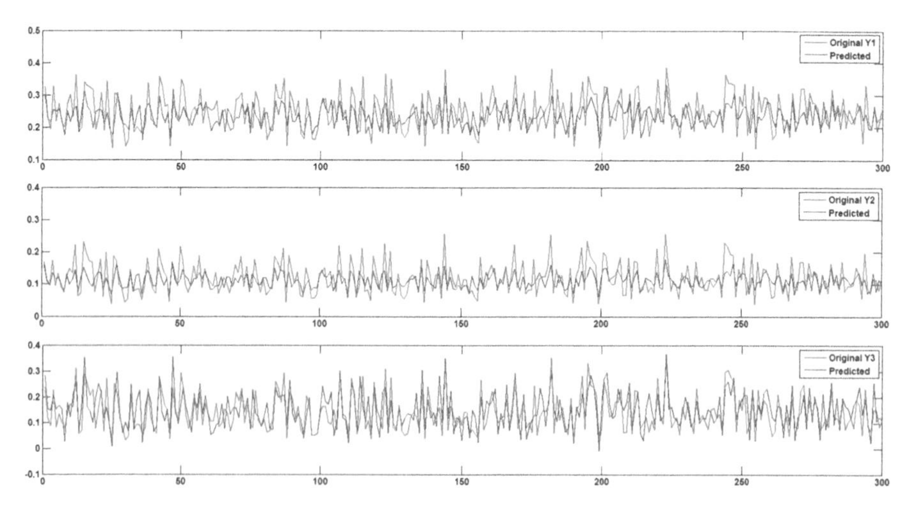

Fig. 3.15 Testing data (target) used to demonstrate iterative (steepest-descent) based regression technique

7. Repeat 3 and 4 until the number of epochs equals to N.
8. Declare the latest $\mathbf{W}$ as the solution.

Illustrations on linear regression using iterative methods are given in Figs. 3.14 and 3.15.

regression-sequential.m

```
%Solving regression using iterative technique
%Linear regression solved using pseudo inverse
%Data generation with noise
W1=rand(10,5);
B1=rand(5,1);
W2=rand(5,3);
B2=rand(3,1);
INPUT=rand(1000,10)*2-1;
H=INPUT*W1+repmat(B1',1000,1);
H=1./(1+exp(H))
OUTPUT=H*W2+repmat(B2',1000,1);
OUTPUT=1./(1+exp(OUTPUT));
INPUTTRAIN=INPUT(1:1:400,:);
INPUTVAL=INPUT(401:1:700,:);
INPUTTEST=INPUT(701:1:1000,:);
OUTPUTTRAIN=OUTPUT(1:1:400,:)+randn(400,3)*0.2;
OUTPUTVAL=OUTPUT(401:1:700,:)+randn(300,3)*0.2;
OUTPUTTEST=OUTPUT(701:1:1000,:);
sigma=0.9:0.01:2;
sigma=1.98;
[V,M]=kmeans(INPUTTRAIN,4);
ERRORV=[];
for iter=1:1:length(sigma)
%Using Gaussian kernel based polynomial regression
[V1,I1]=min(sum((INPUTTRAIN-repmat(M(1,:),
400,1)).^2));
[V2,I2]=min(sum((INPUTTRAIN-repmat(M(2,:),
400,1)).^2));
[V3,I3]=min(sum((INPUTTRAIN-repmat(M(3,:),
400,1)).^2));
[V4,I4]=min(sum((INPUTTRAIN-repmat(M(3,:),
400,1)).^2));
sigma=mean([V1 V2 V3 V4]);
s=sigma(iter);
B1=[];
B2=[];
B3=[];
B4=[];
for i=1:1:size(INPUTTRAIN,1)
B1=[B1 gbf(INPUTTRAIN(i,:),M(1,:),s)];
B2=[B2 gbf(INPUTTRAIN(i,:),M(2,:),s)];
B3=[B3 gbf(INPUTTRAIN(i,:),M(3,:),s)];
B4=[B4 gbf(INPUTTRAIN(i,:),M(4,:),s)];
end
```

```
%Formulation of matrix
MAT=[ones(400,1) B1' B2' B3' B4'];
W=rand(5,3);
ERROR=[];
for epoch=1:1:100
    for data=1:1:length(MAT)
        temp=MAT(data,:)*W;
         W(:,1)=W(:,1)-0.01*(temp(1)-OUTPUTTRAIN
         (data,1))*MAT(data,:)';
         W(:,2)=W(:,2)-0.01*(temp(2)-OUTPUTTRAIN
         (data,2))*MAT(data,:)';
         W(:,3)=W(:,3)-0.01*(temp(3)-OUTPUTTRAIN
         (data,3))*MAT(data,:)';
     end
    ERROR=[ERROR sum((temp-OUTPUTTRAIN(data,:)).^2)];
end
%VALLIDATION
B1=[];
B2=[];
B3=[];
B4=[];
for i=1:1:size(INPUTVAL,1)
B1=[B1 gbf(INPUTVAL(i,:),M(1,:),s)];
B2=[B2 gbf(INPUTVAL(i,:),M(2,:),s)];
B3=[B3 gbf(INPUTVAL(i,:),M(3,:),s)];
B4=[B4 gbf(INPUTVAL(i,:),M(4,:),s)];
end
MAT=[ones(300,1) B1' B2' B3' B4'];
OUTPUTVALR=MAT*W;
ERRORV=[ERRORV sum(sum((OUTPUTVAL-OUTPUTVALR).^2))];
end
[V,I]=min(ERRORV);
s=sigma(I);
B1=[];
B2=[];
B3=[];
B4=[];
ERRORTEST=[];
for i=1:1:size(INPUTTEST,1)
B1=[B1 gbf(INPUTTEST(i,:),M(1,:),s)];
B2=[B2 gbf(INPUTTEST(i,:),M(2,:),s)];
B3=[B3 gbf(INPUTTEST(i,:),M(3,:),s)];
B4=[B4 gbf(INPUTTEST(i,:),M(4,:),s)];
end
MAT=[ones(300,1) B1' B2' B3' B4'];
```

```
OUTPUTTESTR=MAT*W;
figure
subplot(3,1,1)
plot(OUTPUTTEST(:,1),'r')
hold on
plot(OUTPUTTESTR(:,1),'b')
subplot(3,1,2)
plot(OUTPUTTEST(:,2),'r')
hold on
plot(OUTPUTTESTR(:,2),'b')
subplot(3,1,3)
plot(OUTPUTTEST(:,3),'r')
hold on
plot(OUTPUTTESTR(:,3),'b')
ERRORTEST=[ERRORTEST sum(sum((OUTPUTTEST-OUTPUTTESTR).
^2))];

function [res]=gbf(i,m,s)
res=exp(-((i-m)*(i-m)')/(2*s^2));
```

3.9 Kernel Trick for Regression

The mean matrix mentioned in the Bayes technique is represented as $\mathbf{M} = (\alpha I + \beta F^T F)^{-1} \beta F^T T$. The output vector corresponding to the arbitrary vector $\mathbf{x}$ is obtained as $(\mathbf{M})^T f(\mathbf{x})$. Note that the size of the matrix $\mathbf{M}$ is $d \times r$ (r is the number of basis functions) and the size of the vector $f(\mathbf{x})$ is $\mathbf{r} \times 1$.

The output vector corresponding to the arbitrary vector $\mathbf{x}$ is written as follows:

$$f(\mathbf{x})^T \mathbf{M} \tag{3.46}$$

$$= f(\mathbf{x})^T (\alpha I + \beta \mathbf{F}^T \mathbf{F})^{-1} \beta \mathbf{F}^T \mathbf{T} \tag{3.47}$$

$$\sum_{k=1}^{k=d} f(\mathbf{x})^T (\alpha I + \beta \mathbf{F}^T \mathbf{F})^{-1} \beta f(\mathbf{x_k})^T \mathbf{t_k^T} \tag{3.48}$$

where $\mathbf{t_k^T}$ is the kth row of the matrix $\mathbf{T}$. It is noted that $f(\mathbf{x})^T (\alpha I + \beta F^T F)^{-1} \beta f(\mathbf{x_k})^T$ is the scalar and is the function of $\mathbf{x}$ and $\mathbf{x_k}$. Instead of formulating the function f, we can formulate the kernel function as $\mathbf{k}(\mathbf{x}, \mathbf{x_k})$. Rewriting (3.48), we get the following:

$$\sum_{k=1}^{k=d} k(\mathbf{x}, \mathbf{x_k}) \mathbf{t_k} \tag{3.49}$$

Given $\mathbf{x}$, the corresponding output $y(\mathbf{x})$ is obtained as the linear combinations of kernel functional values computed for the individual training data scaled with the corresponding target values t_k. This is popularly called the "Kernel-Trick." The steps involved in formulating the linear regression using kernel trick are summarized as follows:

1. Let the collected input data be $\mathbf{x_1}, \ldots, \mathbf{x_{100}}$ and the corresponding output data be $\mathbf{t_1}, \ldots, \mathbf{t_{100}}$. These 100 data are divided into training and validation data.
2. Training data: Input vectors $\mathbf{x_1}, \mathbf{x_2}, \ldots, \mathbf{x_{70}}$ and the corresponding target vectors $\mathbf{t_1}, \mathbf{t_2}, \ldots, \mathbf{t_{70}}$; validation data: Input vectors $\mathbf{x_{71}}, \mathbf{x_{72}}, \ldots, \mathbf{x_{100}}$ and the corresponding target vectors $\mathbf{t_{71}}, \mathbf{t_{72}}, \ldots, \mathbf{t_{100}}$.
3. Compute $\mathbf{t_u^{est}} = \sum_{k=1}^{k=70} k(\mathbf{x_k}, \mathbf{x_u})\mathbf{t_k}$ using the particular tuning parameter σ, (which is used in the kernel function) for $u = 71 \cdots 100$.
4. Compute the sum squared error as $\mathrm{SSE} = \sum_{m=71}^{m=100} (t_m - t_m^{est})^2$.
5. Repeat 3 and 4 for various values of σ and identify the σ that gives the minimum value of SSE.
6. Use the tuning parameter to model the linear regression model as $y(\mathbf{x}) = \sum_{k=1}^{k=70} k(\mathbf{x_k}, \mathbf{x})\mathbf{t_k}$.
7. It is noted that the tuning parameter is used to compute the kernel function. Figures 3.16 and 3.17 are the input and output data obtained using kernel-based regression techniques. It is seen that the actual output and the estimated output are almost identical.

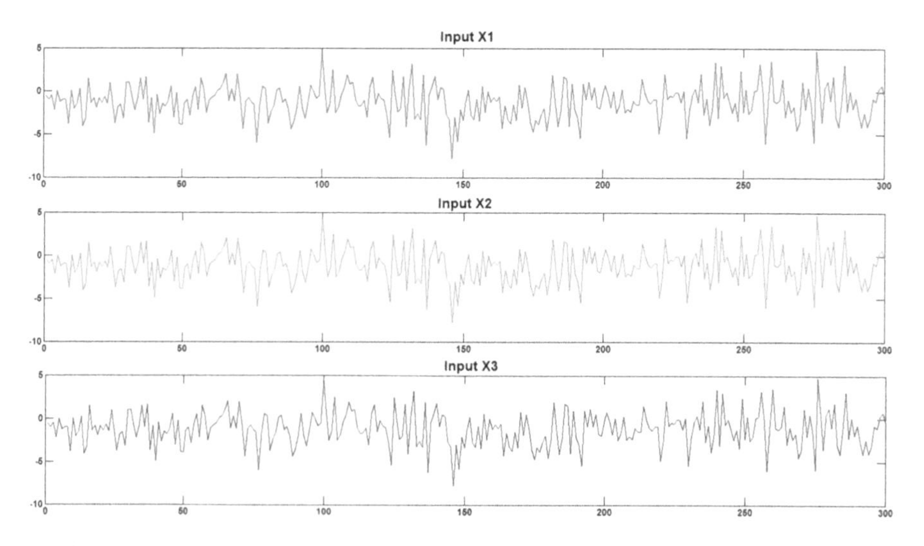

Fig. 3.16 Input data used to demonstrate kernel-based regression technique

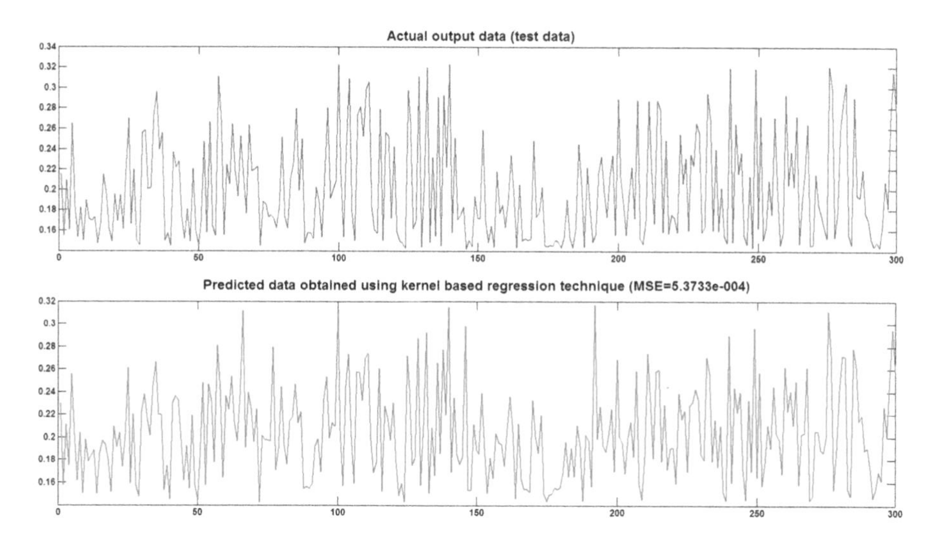

Fig. 3.17 Output data used to demonstrate kernel-based regression technique

3.10 Bayes Linear Regression by Estimating α and β

3.10.1 Review of the Bayes Linear Regression

The Bayes technique for Linear regression model is summarized as follows:

1. Given the noisy observation $\mathbf{t_n} = y(\mathbf{x_n}) + \epsilon$ $(n = 1 \cdots N)$, $y(\mathbf{x_n})$ is modeled as $\mathbf{w}^T\phi(\mathbf{x_n})$.
2. The requirement is to obtain $\mathbf{w}$. The optimal solution for $\mathbf{w}$ is the conditional mean $E(\mathbf{w}/\mathbf{t}, \mathbf{x})$
3. The posterior density function $p(\mathbf{w}/\mathbf{t}, \mathbf{x})$ is obtained using the likelihood function $p(\mathbf{t}/\mathbf{x}, \mathbf{w})$ and the prior density function using the following relationship: $p(\mathbf{w}/\mathbf{t}, \mathbf{x}) = \frac{p(\mathbf{t}/\mathbf{x},\mathbf{w})p(\mathbf{w})}{p(\mathbf{t})}$. It is assumed that the prior density function of $\mathbf{w}$, i.e., $p(\mathbf{w})$ is modeled as Gaussian with mean $\mathbf{m_o}$ and co-variance matrix $\mathbf{S_0}$. As $p(\mathbf{t}/\mathbf{x}, \mathbf{w})$ is the likelihood function with mean vector $\mathbf{w}^T\phi(\mathbf{x_n})$ and co-variance matrix $\frac{1}{\beta}I$, we get the posterior density function $p(\mathbf{w}/\mathbf{t}, \mathbf{x})$, which is also Gaussian with mean vector $\mathbf{m_N}$ and co-variance matrix $\mathbf{S_N} = (\mathbf{S_0^{-1}} + \beta\Phi^T\Phi)^{-1}$. They are obtained as $\mathbf{m_N} = \mathbf{S_N}(\mathbf{S_0^{-1}}\mathbf{m_0} + \beta\Phi^T\mathbf{t})$.
4. As the posterior density function is Gaussian distributed, the conditional mean, the conditional median, and the MAP estimate of the random vector are the mean of the posterior density function.
5. Thus the Bayes estimation of the vector $\mathbf{w}$ is given as $\mathbf{m_N} = \mathbf{S_N}(\mathbf{S_0^{-1}}\mathbf{m_0}+\beta\Phi^T\mathbf{t})$, where $\boldsymbol{\Phi}$ is the matrix with each row is the vector $\phi(x_n)^T$. If M basis functions are used, $\boldsymbol{\Phi}$ is the matrix with size $N \times M$.

6. For the special case when $\mathbf{m_0} = 0$ and $\mathbf{S_0} = \frac{1}{\alpha}\mathbf{I}$, $\mathbf{m_N}$ is given as $\beta \mathbf{S_N} \Phi^T t = \beta(\alpha \mathbf{I} + \beta \Phi^T \Phi)^{-1} \Phi^T t$.
7. The requirement is to tune the best solutions for the variables α and β. This is done using evidence approximation as described in the following section.

3.10.2 Evidence Approximation

The prediction probability density function $p(\mathbf{t_{test}}/\mathbf{x_{test}}, \mathbf{t_{train}}, \alpha, \beta, w)$ is the density function associated with the random variable t_{test} obtained as the output of the trained Bayes model (described by α,β, and $\mathbf{w}$) given the input vector $\mathbf{x_{test}}$. The prediction pdf is written as follows:

$$p(\mathbf{t_{test}}/\mathbf{x_{test}}, \mathbf{t_{train}}, \alpha, \beta, w) = \int\int\int p(\mathbf{t_{test}}/\mathbf{w}, \beta) p(\mathbf{w}/\mathbf{t_{train}}, \alpha, \beta) \times p(\alpha\beta/\mathbf{t_{train}}) d\mathbf{w} d\alpha d\beta$$

$p(\mathbf{t_{test}}/\mathbf{w}, \beta)$ is Gaussian distributed with mean vector $\mathbf{w}^T \boldsymbol{\phi}(\mathbf{x})$ and co-variance matrix $\frac{1}{\beta}\mathbf{I}$. Also $p(\mathbf{w}/\mathbf{t_{train}}, \alpha, \beta)$ is Gaussian distributed with mean vector $\mathbf{m_N}$ and co-variance matrix $\mathbf{S_N}$. The prediction distribution depends on the optimal choice of $\mathbf{w}$ by maximizing $p(\mathbf{w}/\mathbf{t_{train}}, \alpha, \beta)$ for the fixed α and β. In the same fashion, the best choice of α and β is chosen by maximizing $p(\alpha\beta/\mathbf{t_{train}})$ as α_o and β_o. Using this $\mathbf{w_o}$ is obtained as $\beta_o(\alpha_o I + \beta \Phi^T \Phi)^{-1} \Phi^T t$, which is further used to obtain the $(p(\mathbf{t_{test}}/\mathbf{w_o}, \beta_o))$ as Gaussian distributed with mean $\mathbf{w_o}^T \phi(\mathbf{x_{test}})$ and the co-variance matrix $\frac{1}{\beta}\mathbf{I}$. Thus the best choice for α and β obtained by maximizing the function $p(\alpha, \beta/\mathbf{t_{train}})$ is described in the following section.

3.10.3 Solving α and β for Linear Regression Using Bayes Technique

1. $p(\alpha, \beta/\mathbf{t_{train}}) = \frac{p(\mathbf{t_{train}}/\alpha\beta) p(\alpha\beta)}{p(\mathbf{t_{train}})}$. If $p(\alpha\beta)$ is uniform distributed, $p(\alpha, \beta/\mathbf{t_{train}})$ is maximized by maximizing $p(\mathbf{t_{train}}/\alpha\beta)$.
2. $p(\mathbf{t_{train}}/\alpha\beta) = \int p(\mathbf{t_{train}}/\mathbf{w}, \beta) p(\mathbf{w}/\alpha) d\mathbf{w}$.
3. $\int p(\mathbf{t_{train}}/\mathbf{w}, \beta)$ is the likelihood function which is Gaussian distributed with mean $\mathbf{w}^T \phi(\mathbf{x_n})$ and co-variance matrix $\frac{1}{\beta}\mathbf{I}$, and $p(\mathbf{w}/\alpha)$ is Gaussian distributed with mean zero and co-variance matrix $\frac{1}{\alpha}\mathbf{I}$.

4. Thus $ln(p(\mathbf{t_{train}}/\alpha\beta))$ is simplified as $\frac{M}{2}ln(\alpha) + \frac{N}{2}ln(\beta) - E(\mathbf{m_N}) - ln|\mathbf{A}| - \frac{N}{2}ln(2\pi)$, where

$$E(\mathbf{w}) = \frac{\beta}{2}||\mathbf{t} - \Phi\mathbf{w}||^2 + \frac{\alpha}{2}\mathbf{w}^T\mathbf{w} \tag{3.50}$$

$$E(\mathbf{m_N}) = \frac{\beta}{2}||\mathbf{t} - \Phi\mathbf{m_N}||^2 + \frac{\alpha}{2}\mathbf{m_N}^T\mathbf{m_N} \tag{3.51}$$

$$\mathbf{A} = \Delta\Delta E(\mathbf{w}) = \alpha I + \beta\phi^T\phi \tag{3.52}$$

$$\mathbf{m_N} = \beta\mathbf{A}^{-1}\phi^T\mathbf{t} \tag{3.53}$$

3.10.4 Maximizing $p(\mathbf{t_{train}}/\alpha\beta)$ Using E–M Algorithm

- Expectation stage:
  1. Initialize α and β
  2. Compute the eigenvector $\mathbf{u_i}$ and the corresponding eigenvalues λ_i of the matrix

$$(\beta\Phi^T\Phi)\mathbf{u_i} = \lambda_i\mathbf{u_i} \tag{3.54}$$

  3. Compute $\mathbf{m_N} = \beta\mathbf{A}^{-1}\phi^T\mathbf{t}$, where $\mathbf{A} = \alpha\mathbf{I} + \beta\Phi^T\Phi$
  4. Compute $\gamma = \sum_i \frac{\lambda_i}{\alpha+\lambda_i}$
- Maximization stage:
  1. Compute $\alpha = \frac{\gamma}{\mathbf{m_N}^T\mathbf{m_N}}$
  2. Compute $\beta = (\frac{1}{N-\gamma}\sum_{n=1}^{n=N}(\mathbf{t_n} - \mathbf{m_N}\phi(\mathbf{x_n}))^2)^{-1}$

Repeat Expectation and Maximization state for finite number of states until it gets converged. Note: Given α, β, and $\mathbf{w}$, the prediction distribution is represented as follows:

$$\int\int\int p(\mathbf{t_{test}}/\mathbf{w}, \beta)p(\mathbf{w}/\mathbf{t_{train}}, \alpha, \beta)p(\alpha\beta/\mathbf{t_{train}})d\mathbf{w}d\alpha d\beta \tag{3.55}$$

The prediction distribution is also Gaussian distributed with mean vector $\mathbf{m_N}^T\phi(x)$ and variance $\frac{1}{\beta} + \phi(\mathbf{x})^T\mathbf{S_N}\phi(\mathbf{x})$. The linear regression based on the estimation of α and β in Bayes regression is illustrated in Figs. 3.18 and 3.19.

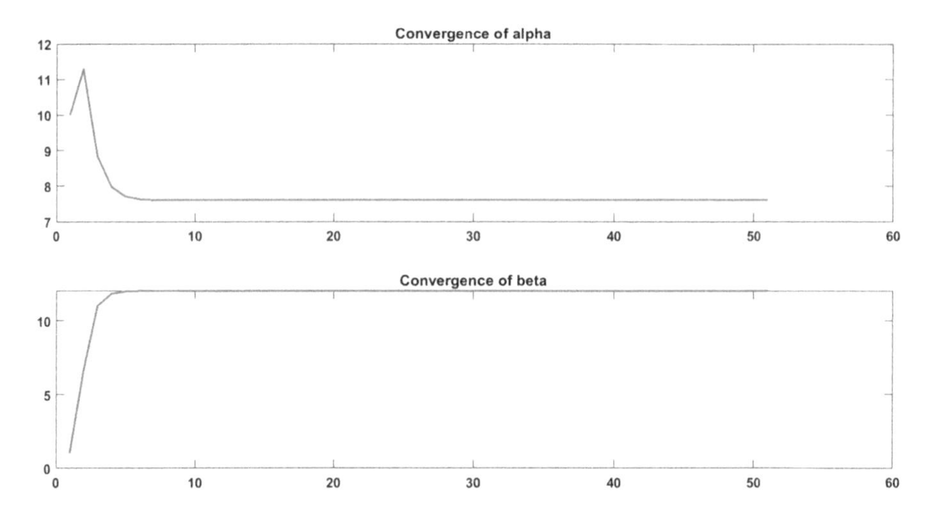

Fig. 3.18 Illustration of convergence of α and β in Bayes regression

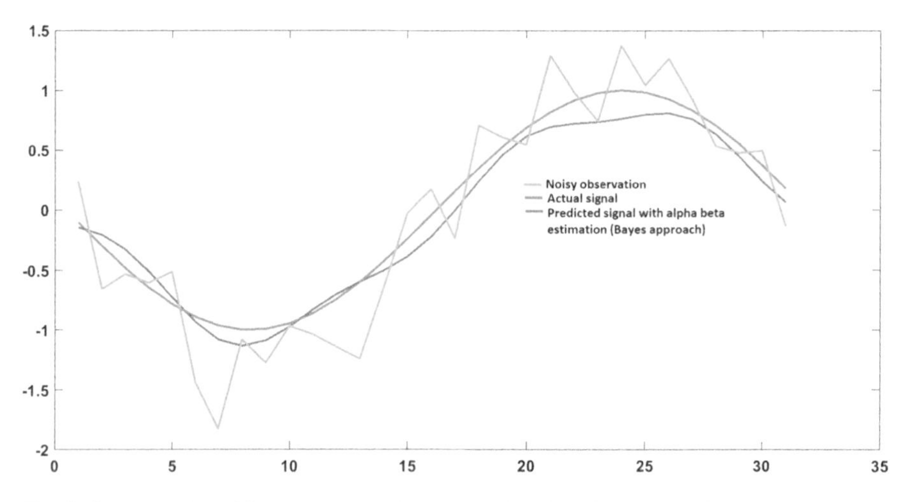

Fig. 3.19 Illustration of linear regression based on estimation of α and β in Bayes regression

Bayesregression.m

```
%Bayes regression
%Training set
v=0.5;
x=-pi:0.1:pi;
beta=10;
t=sin(x)+randn(1,length(x))*(1/sqrt(beta));
TRAINx=x(1:2:length(x));
TRAINt=t(1:2:length(x));
```

```
TESTx=x(2:2:length(x));
TESTt=t(2:2:length(x));
%Estimate alpha and beta
%Let the number of basis function (Gaussian) be 20
%and mean of the basis function is uniformly spaced
between
%-pi to pi
MEAN=linspace(-pi,pi,20);
%Construncting the matrix PHI
PHI=[];
for i=1:1:length(TRAINx)
    temp=[1];
    for j=1:1:length(MEAN)
temp=[temp gbf(TRAINx(i),MEAN(j),v)];
    end
    PHI=[PHI;temp];
end
%Initialize alpha and beta
alpha=10;
beta=1;
Alphacol=[alpha];
Betacol=[beta];
for iteration=1:1:50
[E,D]=eig(beta*PHI'*PHI);
A=(alpha*eye(length(MEAN)+1)+beta*PHI'*PHI);
mN=beta*inv(A)*PHI'*TRAINt';
gamma=0
for i=1:1:size(D,2)
    gamma=gamma+D(i,i)/(alpha+D(i,i));
end
alpha=gamma/(mN'*mN);
s=0;
for i=1:1:length(TRAINx)
    s=s+sum((TRAINt(i)-mN'*PHI(i,:)').^2)
end
beta=inv((1/(length(TRAINx)-gamma))*s);
Alphacol=[Alphacol alpha];
Betacol=[Betacol beta];
end
%Final mN
A=(alpha*eye(length(MEAN)+1)+beta*PHI'*PHI);
mN=beta*inv(A)*PHI'*TRAINt';
%Check with test data
PHItest=[];
for i=1:1:length(TESTx)
```

```
        temp=[1];
        for j=1:1:length(MEAN)
    temp=[temp gbf(TESTx(i),MEAN(j),v)];
        end
        PHItest=[PHItest;temp];
    end
    y_predict=PHItest*mN;
    y_actual=sin(x(2:2:length(x)));
    plot(y_predict)
    hold on
    plot(y_actual)
    plot(TESTt)
    figure
    subplot(2,1,1)
    plot(Alphacol)
    title('Convergence of alpha')
    subplot(2,1,2)
    plot(Betacol)
    title('Convergence of beta')
```

3.10.5 Demonstration on Prediction Distribution

The predictive distribution conditioned on the training set α, β, and w is as follows:

$$\int\int\int p(\mathbf{t_{test}}/\mathbf{w},\beta)p(\mathbf{w}/\mathbf{t_{train}},\alpha,\beta)p(\alpha\beta/\mathbf{t_{train}})d\mathbf{w}d\alpha d\beta \tag{3.56}$$

By choosing the proper choice for α and β, the predictive distribution is obtained as the Gaussian distributed with mean vector $\mathbf{m_N}^T\phi(x)$ and variance $\frac{1}{\beta}+\phi(x)^T\mathbf{S_N}\phi(\mathbf{x})$, where

$$\mathbf{m_N}=\beta(\alpha\mathbf{I}+\beta\Phi^T\Phi)^{-1}\Phi^T\mathbf{t_{train}} \tag{3.57}$$

For the given training dataset, the optimal α and β along with $\mathbf{m_N}$ are obtained and are used to obtain the outcome $\mathbf{t}=\mathbf{m_N}^T\phi(\mathbf{x_{test}})+\epsilon$ corresponding to the test input data $\mathbf{x_{test}}$, where ϵ is the additive Gaussian noise with variance $\frac{1}{\beta}$. The ensemble of outcomes as the function of $\mathbf{x_{test}}$ (ranging from $-\pi$ to π) is illustrated in Fig. 3.20. The variance of the outcome for every test input $\mathbf{x_{test}}$ is estimated and is computed to the results obtained using (3.58). It is found that as the number of basis function is increasing, the variance is getting minimized. After a few iterations, the variance value mainly depends upon the factor $\frac{1}{\beta}$ (refer Fig. 3.21).

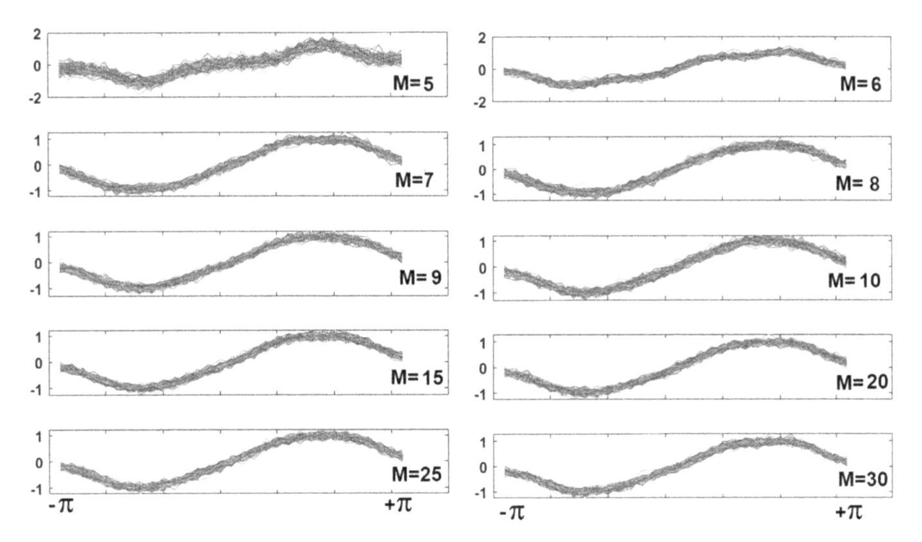

Fig. 3.20 Ensemble of outcomes $\mathbf{t} = \mathbf{m_N}^T\mathbf{x_{test}} + \epsilon$ as the function of $\mathbf{x_{test}}$ with M number of basis functions

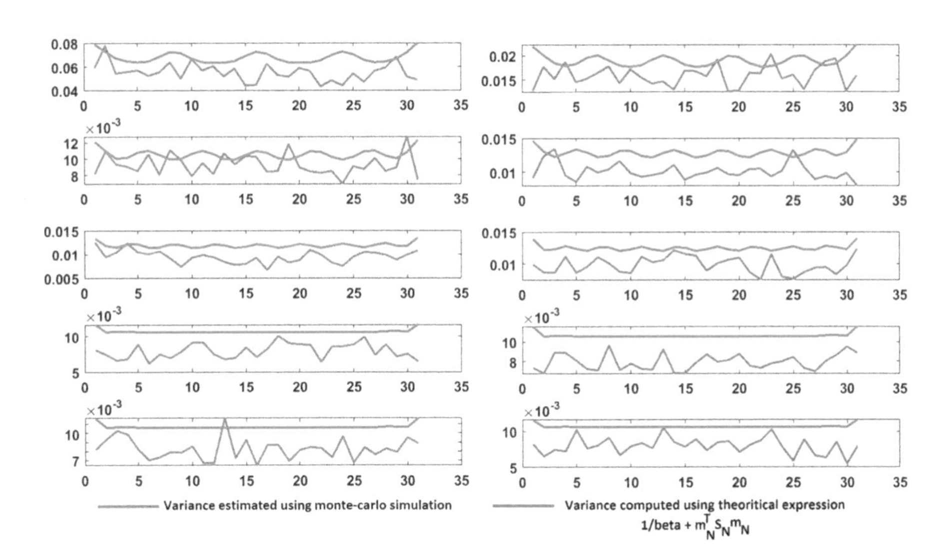

Fig. 3.21 Comparison of variance estimated using Monte-Carlo simulation and using theoretical expression with M number of basis functions

predictivedistribution.m

```
%Prediction distribution
v=0.5;
x=-pi:0.1:pi;
betaactual=100;
t=sin(x)+randn(1,length(x))*(1/sqrt(betaactual));
TRAINx=x(1:2:length(x));
TRAINt=t(1:2:length(x));
%Estimate alpha and beta
%Let the number of basis function (Gaussian) be 20
%and mean of the basis function is uniformly spaced
between
%-pi to pi
u=1;
bn=[5 6 7 8 9 10 15 20 25 30];
for attempt=1:1:10;
TPREDICT=[];
mNCOL=[];
MEAN=linspace(-pi,pi,bn(attempt));
PHI=[];
for i=1:1:length(TRAINx)
temp=[1];
   for j=1:1:length(MEAN)
temp=[temp gbf(TRAINx(i),MEAN(j),v)];
    end
    PHI=[PHI;temp];
end
%Initialize alpha and beta
alpha=10;
beta=1;
Alphacol=[alpha];
Betacol=[beta];
for iteration=1:1:50
[E,D]=eig(beta*PHI'*PHI);
A=(alpha*eye(length(MEAN)+1)+beta*PHI'*PHI);
mN=beta*inv(A)*PHI'*TRAINt';
gamma=0
for i=1:1:size(D,2)
    gamma=gamma+D(i,i)/(alpha+D(i,i));
end
alpha=gamma/(mN'*mN);
s=0;
for i=1:1:length(TRAINx)
    s=s+sum((TRAINt(i)-mN'*PHI(i,:)').^2)
end
```

```
beta=inv((1/(length(TRAINx)-gamma))*s);
Alphacol=[Alphacol alpha];
Betacol=[Betacol beta];
end
%Final mN
A=(alpha*eye(length(MEAN)+1)+beta*PHI'*PHI);
mN=beta*inv(A)*PHI'*TRAINt';

for i=1:1:100
x=-pi:0.1:pi;
betaactual=100;
t=sin(x)+randn(1,length(x))*(1/sqrt(betaactual));
TESTx=x(2:2:length(x));
TESTt=t(2:2:length(x));
%Construncting the matrix PHItest
PHItest=[];
for i=1:1:length(TESTx)
    temp=[1];
    for j=1:1:length(MEAN)
temp=[temp gbf(TESTx(i),MEAN(j),v)];
    end
    PHItest=[PHItest;temp];
end
t_predict=PHItest*mN+randn(length(TESTx),1)*(1/
sqrt(beta));
TPREDICT=[TPREDICT t_predict];
end
figure(1)
subplot(5,2,u)
plot(TPREDICT)
figure(2)
subplot(5,2,u)
plot(var(TPREDICT'))
s1=[];
SN=inv(alpha*eye(size(PHI,2))+beta*PHI'*PHI);
for j=1:1:size(PHItest,1)
s1=[s1 (1/beta)+PHItest(j,:)*SN*PHItest(j,:)'];
end
hold on
plot(s1,'r')
u=u+1;
end
```

3.11 Relevance Vector Machine (RVM)

The classical SVM problem (refer Chap. 2) of the form $y(\mathbf{x}) = \sum_{n=1}^{n=N} \mathbf{w}_n k(\mathbf{x}, \mathbf{x_n}) + b$ can be viewed as the linear regression problem of the form $y(\mathbf{x}) = \sum_{n=1}^{n=N} \mathbf{w_n} \phi(\mathbf{x_n})$. This can be solved using Bayes based technique (refer Sect. 3.10) with the assumption that the prior density function is Gaussian distributed with mean zero vector and diagonal co-variance matrix ($\mathbf{S_0}$) with diagonal elements filled up with $[\frac{1}{\alpha_1}\ \frac{1}{\alpha_2} \cdots \frac{1}{\alpha_N}]$ as described below.

3.11.1 Review of the Bayes Linear Regression

The Bayes technique for linear regression model is summarized as follows:

1. Given the noisy observation $\mathbf{t_n} = y(\mathbf{x_n}) + \epsilon$ $(n = 1 \cdots N)$, $y(\mathbf{x}_n)$ is modeled as $\mathbf{w}^T \phi(\mathbf{x}_n)$.
2. The requirement is to obtain $\mathbf{w}$. The optimal solution for $\mathbf{w}$ is the conditional mean $E(\mathbf{w}/\mathbf{t}, \mathbf{x})$.
3. The posterior density function $p(\mathbf{w}/\mathbf{t}, \mathbf{x})$ is obtained using the likelihood function $p(\mathbf{t}/\mathbf{x}, \mathbf{w})$ and the prior density function using the following relationship: $p(\mathbf{w}/\mathbf{t}, \mathbf{x}) = \frac{p(\mathbf{t}/\mathbf{x},\mathbf{w})p(\mathbf{w})}{p(\mathbf{t})}$. It is assumed that the prior density function of $\mathbf{w}$, i.e., $p(\mathbf{w})$ is modeled as Gaussian with mean $\mathbf{m_o}$ and co-variance matrix $\mathbf{S_0}$. As $p(\mathbf{t}/\mathbf{x}, \mathbf{w})$ is the likelihood function with mean vector $\mathbf{w}^T \phi(\mathbf{x_n})$ and co-variance matrix $\frac{1}{\beta}\mathbf{I}$, we get the posterior density function $p(\mathbf{w}/\mathbf{t}, \mathbf{x})$, which is also Gaussian with mean vector $\mathbf{m_N}$ and co-variance matrix $\mathbf{S_N} = (\mathbf{S_0}^{-1} + \beta \Phi^T \Phi)^{-1}$. They are obtained as $\mathbf{m_N} = \mathbf{S_N}(\mathbf{S_0}^{-1}\mathbf{m_0} + \beta \Phi^T t)$.
4. As the posterior density function is Gaussian distributed, the conditional mean, the conditional median, and the MAP estimate of the random vector are the mean of the posterior density function.
5. Thus the Bayes estimation of the vector $\mathbf{w}$ is given as $\mathbf{m_N} = \mathbf{S_N}(\mathbf{S_0}^{-1}\mathbf{m_0} + \beta \Phi^T \mathbf{t})$, where Φ is the matrix with each row is the vector $(\phi(\mathbf{x_n}))^T$. If M basis functions are used, Φ is the matrix with size $N \times M$.
6. For the special case when $\mathbf{m_0} = 0$ and $\mathbf{S_0}$ with diagonal elements $[\alpha_1\ \alpha_2 \cdots \alpha_N]$, $\mathbf{m_N}$ is given as $\beta \mathbf{S_N} \Phi^T t = \beta(\mathbf{S_0} + \beta \Phi^T \Phi)^{-1} \Phi^T t$.
7. The requirement is to tune the best solutions for the variables $\alpha_i^{\prime} s$ and β. This is done using evidence approximation (by maximizing the marginal likelihood function).

The predictive distribution $p(\mathbf{t_{test}}/\mathbf{t_{train}}, \mathbf{x_{train}}, \mathbf{x_{test}}, \alpha, \beta, w)$ is the likelihood function of the predictive random variable $\mathbf{t_{test}}$ given the test input $\mathbf{x_{test}}$. The expression is rewritten without $\mathbf{x_{test}}$ and $\mathbf{x_{train}}$ as

$$p(\mathbf{t_{test}}/\mathbf{t_{train}}, \alpha, \beta, \mathbf{w}) \tag{3.58}$$

The $p(\mathbf{t_{test}}/\mathbf{t_{train}})$ is obtained marginalizing $p(\mathbf{t_{test}}, \mathbf{w}, \alpha, \beta/\mathbf{t_{train}})$, which is simplified further as $p(\mathbf{t_{test}}/\mathbf{w}, \beta)p(\mathbf{w}/\mathbf{t_{train}}, \alpha, \beta)p(\alpha, \beta/\mathbf{t_{train}})$. The posterior density function $p(\alpha, \beta/\mathbf{t_{train}}) = kp(\mathbf{t}/\alpha\beta)p(\alpha, \beta)$ (where k constant) is maximized to obtain α_m and βm. Using the values obtained (fixed), the predictive distribution $p(\mathbf{t_{test}}/\mathbf{t_{train}})$ is obtained by marginalizing over $\mathbf{w}$. Assuming uniform distribution for α and β, we get $p(\alpha, \beta/\mathbf{t_{train}}) = cp(\mathbf{t_{train}}/\alpha\beta)$, where c is the constant. The pdf $p(\mathbf{t}/\alpha\beta)$ is known as evidence function. The optimal α and β are obtained by maximizing the evidence function.

3.11.2 Summary

1. $p(\mathbf{w}/\alpha)$ is the prior density function with mean zero vector and co-variance matrix $\mathbf{A}$ (with i^{th} diagonal value is α_i) (i^{th} element of the vector α).
2. $p(\mathbf{w}/\mathbf{t})$ is Gaussian distributed with mean vector $\mathbf{m} = \beta\mathbf{B}\Phi^T\mathbf{t}$. $\mathbf{B} = (\mathbf{S_o} + \beta\Phi^T\Phi)^{-1}$.
3. We represent the evidence approximation as $\mathbf{p}(\mathbf{t_{train}}/\alpha\beta) = \int p(\mathbf{t_{train}}/\mathbf{w}, \beta) p(\mathbf{w}/\alpha))d\mathbf{w}$.
4. Thus the evidence approximation is identified as the Gaussian distributed with mean zero and co-variance matrix $\mathbf{C} = \beta^{-1}I + \Phi\mathbf{A}^{-1}\Phi^T$.

   The steps involved in estimating α_i, β are summarized as follows:

1. Initialize α_i, β.
2. Compute $\mathbf{m} = \beta\mathbf{B}\Phi^T\mathbf{t}$, where $\mathbf{B} = (\mathbf{S_0} + \beta\Phi^T\Phi)^{-1}$.
3. Estimate $\gamma_i = 1 - \alpha_i B_{ii}$, where B_{ii} is the ith diagonal element of the matrix $\mathbf{B}$.
4. The new α_i is estimated as $\alpha_i = \frac{\gamma_i}{m_i^2}$.
5. Also β is updated as $\frac{N-\sum_i \gamma_i}{||\mathbf{t}-\Phi\mathbf{m}||^2}$.

In the special case, when α_i's are assumed as identical, we get the following steps:

1. Initialize α and β.
2. Compute $\mathbf{A} = \alpha I + \beta\Phi^T\Phi$.
3. Compute $\mathbf{m_N} = \beta\mathbf{A}^{-1}\Phi^T\mathbf{t}$.
4. Compute the eigenvalue of $\beta\Phi^T\Phi$. Let it be $\lambda_1, \lambda_2, ..., \lambda_M$, where M is the number of basis functions used.
5. Compute $\gamma = \sum_i \frac{\lambda_i}{\alpha+\lambda_i}$.
6. Reestimate $\alpha = \frac{\gamma}{\mathbf{m_N}^T\mathbf{m_N}}$.
7. Reestimate $\beta = \frac{1}{N-\gamma}\sum_{n=1}^{n=N}(\mathbf{t_n} - \mathbf{m_N}\Phi(x_n))^2$.
8. Repeat steps 2–7 until convergence occurs.

The illustration of Relevance Vector Machine is demonstrated in Figs. 3.22 and 3.23.

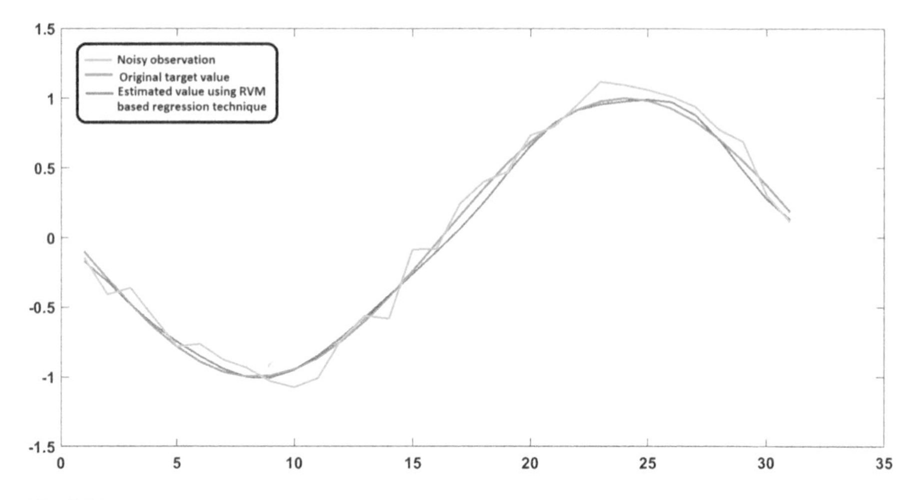

Fig. 3.22 Illustration using Relevance Vector Machine (RVM)

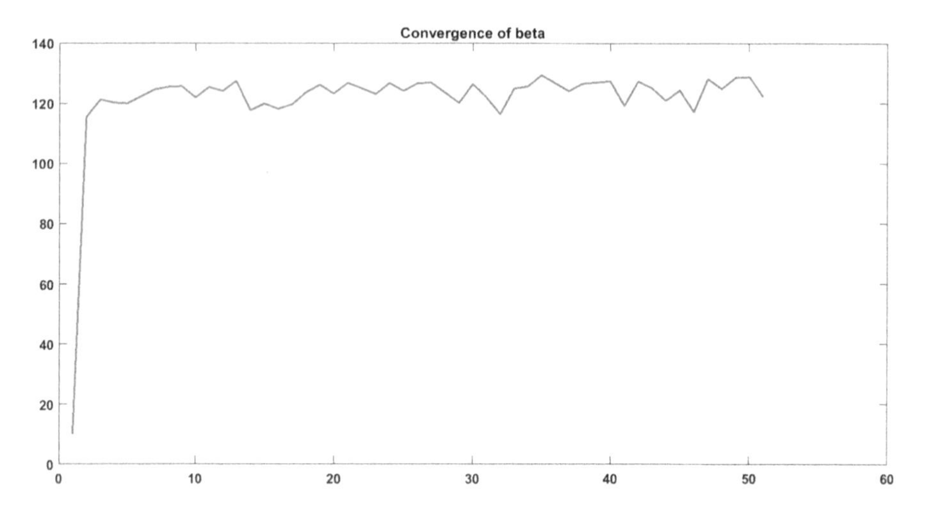

Fig. 3.23 Illustration of convergence of the parameter β in RVM

regression-RVM.m

```
%Relevant Vector Machine
%Bayes regression
%Training set
v=0.5;
x=-pi:0.1:pi;
betaactual=100;
t=sin(x)+randn(1,length(x))*(1/sqrt(betaactual));
TRAINx=x(1:2:length(x));
```

```
TRAINt=t(1:2:length(x));
TESTx=x(2:2:length(x));
TESTt=t(2:2:length(x));
%Estimate alpha and beta
%Let the number of basis function (Gaussian) be 20
%and mean of the basis function is uniformly spaced
between
%-pi to pi
MEAN=linspace(-pi,pi,20);
%Construncting the matrix PHI
PHI=[];
for i=1:1:length(TRAINx)
    temp=[1];
    for j=1:1:length(TRAINx)
temp=[temp gaussiankernel(TRAINx(i),TRAINx(j),v)];
    end
    PHI=[PHI;temp];
end
%Initialize alpha and beta
alpha=rand(1,size(TRAINx,2)+1);
beta=10;
Alphacol{1}=alpha;
Betacol=[beta];
for iteration=1:1:50
A=inv(diag(alpha)+beta*PHI'*PHI);
for i=1:1:33
    gamma(i)=1-alpha(i)*A(i,i);
end
mN=beta*A*PHI'*TRAINt';
for i=1:1:33
    alpha(i)=gamma(i)/(mN(i)^2);
end
s=0;
for i=1:1:length(TRAINx)
    s=s+sum((TRAINt(i)-mN'*PHI(i,:)').^2)
end
beta=inv((1/(length(TRAINx)-sum(gamma)))*s);
Alphacol{iteration}=alpha;
Betacol=[Betacol beta];
end
%Final mN
A=inv(diag(Alphacol{50})+beta*PHI'*PHI);
mN=beta*A*PHI'*TRAINt';
%Check with test data
PHItest=[];
```

```
for i=1:1:length(TESTx)
    temp=[1];
    for j=1:1:length(TRAINx)
temp=[temp gaussiankernel(TESTx(i),TRAINx(j),v)];
    end
    PHItest=[PHItest;temp];
end
y_predict=PHItest*mN;
y_actual=sin(x(2:2:length(x)));
plot(y_predict)
hold on
plot(y_actual)
plot(TESTt)
figure
plot(Betacol)
title('Convergence of beta')
```

3.12 RVM Using Sparse Method for Regression

Thus the evidence approximation is identified as the Gaussian distributed with mean zero and co-variance matrix $\mathbf{C} = \beta^{-1}\mathbf{I} + \Phi\mathbf{A}^{-1}\Phi^T = \beta^{-1}\mathbf{I} + \alpha_1^{-1}\phi_1((x))\phi_1((x))^T + \alpha_2^{-1}\phi_2(x)\phi_2((x))^T + \alpha_3^{-1}\phi_3((x))\phi_3((x))^T + \cdots + \alpha_M^{-1}\phi_M((x))\phi_M((x))^T$, where $\phi_i((x))$ is the ith column of the matrix Φ. Consider for the case when $M = 1$, we get the following: $\mathbf{C} = \beta^{-1}I + \Phi A^{-1}\Phi^T = \beta^{-1} + \alpha_1^{-1}\phi_1(x)\phi_1(x)^T$. The contours of the probability density function of the 2D-Gaussian density function follow ellipse shape. The major and the minor axes of the ellipse are described by its co-variance matrix. In this case, the major axis is in the direction of $\phi_1(\mathbf{x})$. Also if the vector with identical magnitude $\mathbf{t}$ is in the direction of the major axis of the ellipse, the density functional value is maximum. Hence if the observation vector $\mathbf{t}$ is not aligned with the direction of the basis vector $\phi_1(\mathbf{x})$, the probability density value is minimum. Hence α_{11} is chosen such that the second term in the co-variance matrix vanishes. This is achieved by choosing $\alpha_1 a \infty$. Thus the method of selecting the basis vectors that maximize the evidence probability (given training data) is summarized below and hence the corresponding α_i and β are also obtained using the following iterative procedure.

1. Initialize nonzero α_1's and value for β. Also initialize $\alpha_i = \infty$ for all $i \neq 1$. Initialize the $MODEL = [1]$ (the values of the variable $MODEL$ consists of index of the basis functions involved in the regression model).
2. Compute $\mathbf{C}_{-i} = \frac{1}{beta}\mathbf{I} + \sum_{j=1, j\neq i}^{j=M}(1/\alpha_i)\phi_j^T\phi_j$.
3. Compute $s_i = \phi_i(\mathbf{x})\mathbf{C}_{-i}^{-1}\phi_i(\mathbf{x})$.
4. Compute $q_i = \phi_i(\mathbf{x})\mathbf{C}_{-i}^{-1}\mathbf{t}$.

5. If $(q(i)^2 > s(i))$ and $(\alpha_i < \text{inf})$, update $\alpha_i = \alpha_i = \frac{s_i^2}{(q_i^2 - s_i)}$.
6. If $(q(i)^2 > s(i))$ and $(\alpha_i = \text{inf})$, update $\alpha_i = \alpha_i = \frac{s_i^2}{(q_i^2 - s_i)}$. Include i^{th} basis column in the model by updating $MODEL$.
7. If $(q(i)^2 > s(i))$ and $(\alpha_i < \text{inf})$, remove i^{th} basis column from the model by updating $MODEL$.
8. Repeat the steps for all i.
9. Formulate the basis matrix Φ with the inclusion of the basis vectors described by the elements of the matrix $MODEL$.
10. Update the matrix $\mathbf{A} = (diag(\alpha)\mathbf{I} + \beta\Phi^T\Phi)^T$, where $diag(\alpha)$ is the matrix with diagonal elements which are filled up with the elements of the vector $\boldsymbol{\alpha}$.
11. Use the updated Φ matrix to update $\mathbf{m} = \beta\mathbf{A}\Phi^T\mathbf{t}$.
12. Compute $\gamma_i = 1 - \alpha_i A(i, i)$.
13. $\alpha_i = \frac{\gamma_i}{m_i^2}$, where m_i is the ith element of the vector $\mathbf{m}$.
14. Compute $\beta = \frac{N - \sum_i \gamma_i}{||\mathbf{t} - \Phi\mathbf{m}||^2}$.

The illustration of measurement-based Relevance Vector Machine is demonstrated in Figs. 3.24 and 3.25.

regression-RVMsparse.m

```
%Relevant Vector Machine
%(sparsity measurment technique)
%Bayes regression
%Training set
clear all
close all
```

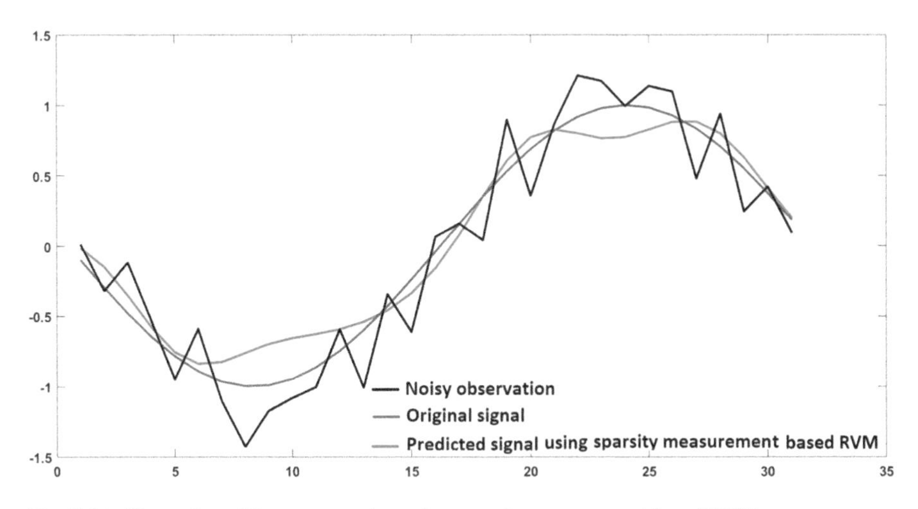

Fig. 3.24 Illustration of linear regression using sparsity measurement based RVM

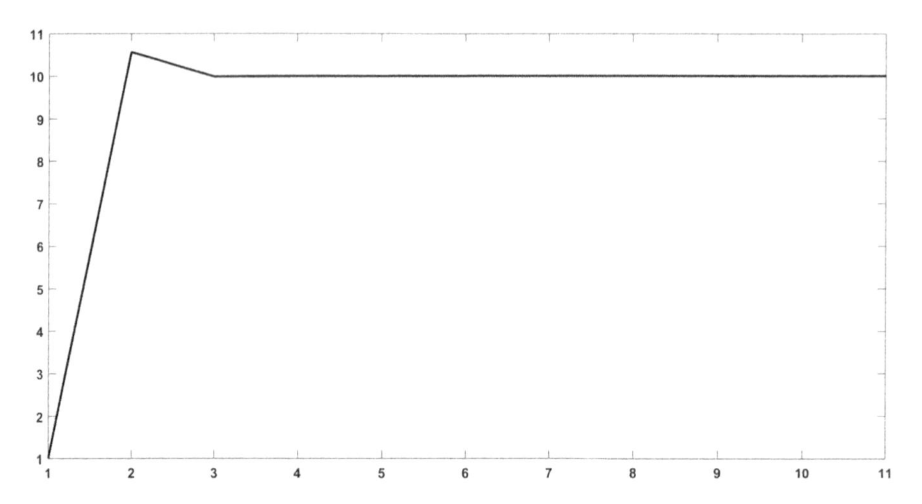

Fig. 3.25 Illustration of convergence of beta in sparsity measurement based RVM

```
v=0.5;
scal=1;
x=-pi:0.1:pi;
betaactual=10;
t=sin(x)+randn(1,length(x))*(1/sqrt(betaactual));
TRAINx=x(1:2:length(x));
TRAINt=t(1:2:length(x))*scal;
TESTx=x(2:2:length(x));
TESTt=t(2:2:length(x))*scal;
%Estimate alpha and beta
%Construncting the matrix PHI
MEAN=linspace(-pi,pi,100);
PHI=[];
for i=1:1:length(TRAINx)
    temp=[1];
    for j=1:1:length(MEAN)
temp=[temp gbf(TRAINx(i),MEAN(j),v)];
    end
    PHI=[PHI;temp];
end
%The alpha vector and beta is solved using the
sparsity analysis as given
%below
beta=1;
alpha=ones(1,size(MEAN,2)+1)*inf;
alpha(1)=rand*0.1;
MODEL=[1];
```

```
Betacol=[beta];
for iteration=1:1:10
for k=1:1:size(MEAN,2)+1
    temp=0;
    for i=1:1:length(size(MEAN,2)+1)
        if(i~=k)
            temp=temp+(1/alpha(i))*PHI(:,i)*PHI(:,i)';
        end
        temp=temp+(1/beta)*eye(size(temp,2));
    end
    s(k)=PHI(:,k)'*inv(temp)*PHI(:,k);
    q(k)=PHI(:,k)'*inv(temp)*TRAINt';
 end
for k=1:1:size(MEAN,2)+1
    if(~(isnan(q(k))|(isnan(s(k)))))
      if(((q(k)^2)>s(k))&(alpha(k)<inf))
alpha(k)=(s(k)^2)/((q(k)^2)-s(k));
elseif((q(k)^2)>s(k))&((alpha(k)==inf))
alpha(k)=(s(k)^2)/((q(k)^2)-s(k));
if(length(find((MODEL-k)==0))==0)
MODEL=[MODEL k];
end
elseif(((q(k)^2)<=s(k))&((alpha(k)<inf)))
alpha(k)=inf;
MODEL=nonzeros(MODEL-k)+k;
MODEL=MODEL';
end
    end
end
gamma=0;
A=0;
A=inv(diag(alpha(MODEL))+beta*PHI(:,MODEL)'*PHI
(:,MODEL));
mN=beta*A*PHI(:,MODEL)'*TRAINt';
for i=1:1:length(MODEL)
     gamma(MODEL(i))=1-alpha(MODEL(i))*A(i,i);
end
s1=sum(([TRAINt- (PHI(:,MODEL)*mN)']).^2);
beta=inv((s1/(length(TRAINx)-sum(gamma))));
Alphacol{iteration}=alpha;
Betacol=[Betacol beta];
end
A=inv(diag(alpha(MODEL))+beta*PHI(:,MODEL)'*PHI
(:,MODEL));
mN=beta*A*PHI(:,MODEL)'*TRAINt';
```

```
%Check with test data
PHItest=[];
for i=1:1:length(TESTx)
    temp=[1];
    for j=1:1:length(MEAN)
temp=[temp gbf(TESTx(i),MEAN(j),v)];
    end
    PHItest=[PHItest;temp];
end
y_predict=PHItest(:,MODEL)*mN;
y_actual=scal*sin(x(2:2:length(x)));
plot(y_predict,'r')
hold on
plot(y_actual,'b')
plot(TESTt,'k')
figure
plot(Betacol,'k')
title('Convergence of beta')
```

3.13 Gaussian Process-Based Regression Techniques

The parametric approach in linear regression involves obtaining the optimal $\mathbf{w}$ such that $t_n = \mathbf{w}^{\mathbf{T}}\phi(x_n)+\epsilon$. In this section, t_n and x_n are assumed as scalar. The extension to the vector is straightforward. The random variable $\boldsymbol{\epsilon}$ is assumed as the multivariate Gaussian density function with mean zero and the co-variance matrix $\frac{1}{\beta}I$. The prior density function of the random vector $\mathbf{w}$ is modeled as Gaussian with mean vector zero and the co-variance matrix $\frac{1}{\alpha}I$. The random vector $\mathbf{y} = \Phi\mathbf{w}$, where

$$\mathbf{y} = [y_1 \ y_2 \ \cdots y_N]^T \tag{3.59}$$

$$\Phi = \begin{bmatrix} \phi_1(x_1) & \phi_2(x_1) & \cdots & \phi_M(x_1) \\ \phi_1(x_2) & \phi_2(x_2) & \cdots & \phi_M(x_2) \\ \cdots & \cdots & \cdots & \cdots \\ \phi_1(x_N) & \phi_2(x_N) & \cdots & \phi_M(x_N) \end{bmatrix} \tag{3.60}$$

The co-variance matrix of the random vector $\mathbf{y}$ is obtained as $\frac{1}{\alpha}K$, where K is the kernel function. The $(m,n)^{\text{th}}$ element of the kernel function is obtained as $\phi(x_m)^T\phi(x_n)$. The random variables $t_1, t_2, ..., t_N$ are assumed as the Gaussian random variables (with mean $w^T\phi(x_k)$ corresponding to t_k and co-variance matrix $\frac{1}{\beta}I$) obtained by tapping across the process at $x_1 \ x_2 \ \cdots \ x_N$. Thus the marginal distribution $p((t_1 \ t_2 \ t_3 \ \cdots \ t_N)^T) = p(\mathbf{t})$ is obtained as follows:

$$p(\mathbf{t}) = \int p(\mathbf{t}/\mathbf{y})p(\mathbf{y})d\mathbf{y} \tag{3.61}$$

The co-variance matrix ($C_{\mathbf{y}}$) and the mean vector ($\mu_{\mathbf{y}}$) of the random vector $\mathbf{t}$ are given as follows:

$$C_{\mathbf{y}} = K + \frac{1}{\beta} I \tag{3.62}$$

Suppose we want to know the distribution of t_{test} corresponding to the input x_{test}, we obtain the joint density function of $p((t_1\ t_2\ t_3\ \cdots\ t_N\ t_{test})^T)$ that follows the Gaussian density function with co-variance matrix as follows:

$$\Phi = \begin{bmatrix} C_y & \mathbf{k} \\ \mathbf{k}^T & k(x_{test}, x_{test}) + \frac{1}{\beta} \end{bmatrix} \tag{3.63}$$

where $\mathbf{k} = [k(x_1, x_{test})\ k(x_2, x_{test})\ \cdots\ k(x_N, x_{test})]^T$. Thus the prediction distribution $p(t_{test}/\mathbf{t})$ is obtained as the Gaussian distributed with mean and co-variance matrix are given by the following:

$$m_{test} = \mathbf{k}^T (C_y)^{-1} \mathbf{t} \tag{3.64}$$

$$v_{test} = \left[k(x_{test}, x_{test}) + \frac{1}{\beta} \right] - \mathbf{k}^T (C_y)^{-1} \mathbf{k} \tag{3.65}$$

We choose m_{test} as the estimate corresponding to x_{test}. This method is identical with the regression method for kernel smoothing, if C_y is chosen as the identity matrix. It is noted that $(m, n)^{\text{th}}$ element of the C_y matrix is $k(x_m, x_n) + \frac{1}{\beta}$. If the noise variance is known, the mean of the prediction distribution $p(t_{test}/\mathbf{t})$ is the regression estimate corresponding to the input x_{test}. Figure 3.26 illustrates regression using Gaussian process (see also Figs. 3.27 and 3.28).

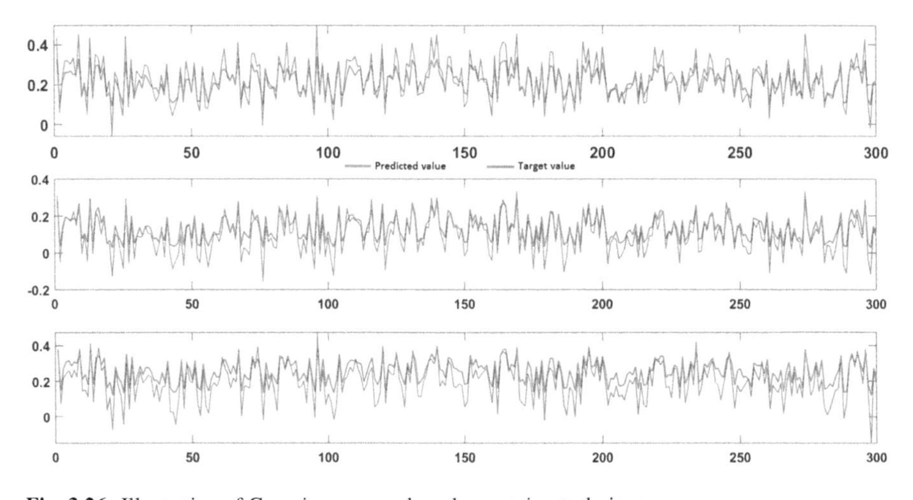

Fig. 3.26 Illustration of Gaussian process-based regression techniques

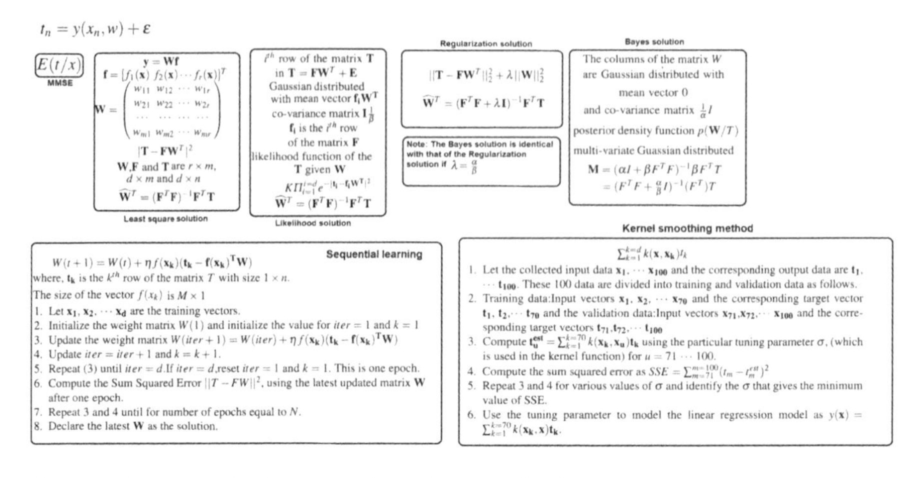

Fig. 3.27 Summary of various regression techniques (Part 1)

$p(t_{test}/x_{test}, t_{train}, \alpha, \beta, w) = \int\int\int p(t_{test}/w, \beta) p(w/t_{train}, \alpha, \beta) p(\alpha\beta/t_{train}) dw d\alpha d\beta$

$p(t_{test}/w, \beta)$ is Gaussian distributed with mean vector $w^T \phi(x)$ co-variance matrix $\frac{1}{\beta} I$

$p(w/t_{train}, \alpha, \beta)$ is Gaussian distributed

BAYES SOLUTION USING EVIDENCE APPROXIMATION

The prediction distribution depends on the optimal choice of w by maximizing $p(w/t_{train}, \alpha, \beta)$ for the fixed α and β

α and β are chosen by maximizing $p(\alpha\beta/t_{train})$ as α_o and β_o

Using this w_o is obtained as $\beta_o(\alpha_o I + \beta \Phi^T \Phi)^{-1} \Phi^T t$.

$(p(t_{test}/w_o, \beta_o)$ as Gaussian distributed with mean $w_o^T \phi(x_{test})$

the co-variance matrix $\frac{1}{\beta} + \phi(x)^T S_N \phi(x)$

Thus the best choice for α and β are obtained by maximizing the func tion $p(\alpha, \beta/t_{train})$ as described below.

1. $p(\alpha, \beta/t_{train}) = \frac{p(t_{train}/\alpha\beta)p(\alpha\beta)}{p(t_{train})}$. If $p(\alpha\beta)$ is uniform distributed, $p(\alpha, \beta/t_{train})$ is maximized by maximizing $p(t_{train}/\alpha\beta)$.
2. $p(t_{train}/\alpha\beta) = \int p(t_{train}/w, \beta) p(w/\alpha) dw$

- Expectation stage:
  1. Initialize α and β
  2. Compute the eigenvector u_i and the corresponding eigenvalues λ_i of the matrix

     $$(\beta \Phi^T \Phi) u_i = \lambda_i u_i$$

  3. Compute $m_N = \beta A^{-1} \phi^T t$, where $A = \alpha I + \beta \Phi^T \Phi$
  4. Compute $\gamma = \sum_i \frac{\lambda_i}{\alpha + \lambda_i}$
- Maximization stage:
  1. Compute $\alpha = \frac{\gamma}{m_N^T m_N}$
  2. Compute $\beta = (\frac{1}{N-\gamma} \sum_{n=1}^{n=N} (t_n - m_N \phi(x_n))^2)^{-1}$

Repeat Expectation and Maximization state for finite number of states until it gets converged.

$p(w/\alpha)$ is the prior density function with mean zero vector and co-variance matrix A (with i^{th} diagonal value is α_i) (i^{th} element of the vector α)

$p(w/t)$ is Gaussian distributed with mean vector $m = \beta B \Phi^T t$, $B = (S_o + \beta \Phi^T \Phi)^{-1}$

We represent the evidence approximation as $p(t_{train}/\alpha\beta) = \int p(t_{train}/w, \beta) p(w/\alpha)) dw$

Thus the evidence approximation is identified as the Gaussian distributed with mean zero and co-variance matrix $C = \beta^{-1} I + \Phi A^{-1} \Phi^T$

RSVM TECHNIQUE

The steps involved in estimating α_i, β are summarized below.

1. Initialize α_i, β.
2. Compute $m = \beta B \Phi^T t$, where $B = (S_0 + \beta \Phi^T \Phi)^{-1}$.
3. Estimate $\gamma_i = 1 - \alpha_i B_{ii}$, where B_{ii} is the ith diagonal element of the matrix B.
4. The new α_i are estimated as $\alpha_i = \frac{\gamma_i}{m_i^2}$.
5. Also β is updated as $\frac{N - \sum_i \gamma_i}{||t - \Phi m||^2}$.

RSVM SPARSITY MEASUREMENT TECHNIQUE

1. Initialize non-zero α_1's and value for β. Also initialize $\alpha_i = \infty$ for all $i \neq 1$. Initialize the $MODEL = [1]$ (The values of the variable $MODEL$ consists of index of the basis functions involved in the regression model)
2. Compute $C_{-i} = \frac{1}{beta} I + \sum_{j=1, j \neq i}^{j=M} (1/\alpha_i) \phi_j^T \phi_j$.
3. Compute $s_i = \phi_i C_{-i}^{-1} \phi_i$
4. Compute $q_i = \phi_i C_{-i}^{-1} t$
5. if $(q(i)^2 > s(i))$ and $(\alpha_i < \inf)$, update $\alpha_i = \alpha_i = \frac{s_i^2}{(q_i^2 - s_i)}$
6. if $(q(i)^2 > s(i))$ and $(\alpha_i = \inf)$, update $\alpha_i = \alpha_i = \frac{s_i^2}{(q_i^2 - s_i)}$. Include i^{th} basis column in the model by updating $MODEL$.
7. if $(q(i)^2 > s(i))$ and $(\alpha_i < \inf)$. Remove i^{th} basis column from the model by updating $MODEL$.
8. Repeat the steps for all i

Fig. 3.28 Summary of various regression techniques (Part 2)

gaussianprocess.m

```
%Gaussian process
W1=rand(10,5);
B1=rand(5,1);
W2=rand(5,3);
B2=rand(3,1);
INPUT=rand(1000,10)*2-1;
H=INPUT*W1+repmat(B1',1000,1);
H=1./(1+exp(H))
OUTPUT=H*W2+repmat(B2',1000,1);
OUTPUT=1./(1+exp(OUTPUT));
```

```
INPUTTRAIN=INPUT(1:1:400,:);
VALIDTRAIN=INPUT(401:1:700,:)
INPUTTEST=INPUT(701:1:1000,:);
OUTPUTTRAIN=OUTPUT(1:1:400,:)+randn(400,3)*0.2;
OUTPUTVALID=OUTPUT(401:1:700,:)+randn(300,3)*0.2;
OUTPUTTEST=OUTPUT(701:1:1000,:);
SSECOL=[];
var=[0.1:0.5:10 10:10:100 100:100:2000];
for iter=1:1:50
%Formulation of kernel matrix
for i=1:1:length(INPUTTRAIN)
    for j=1:1:length(INPUTTRAIN)
        K(i,j)=gbf(INPUTTRAIN(i,:),INPUTTRAIN(j,:),
        var(iter));
    end
end
%Formulation of C matrix
C=K+0.2*eye(400);
%Prediction is obtained as follows
TARGET=[];
for i=1:1:length(VALIDTRAIN)
    temp=[];
    for j=1:1:length(INPUTTRAIN)
temp=[temp gbf(VALIDTRAIN(i,:),INPUTTRAIN(j,:),
var(iter))];
    end
    TARGET=[TARGET (temp*inv(C)*OUTPUTTRAIN)'];
end
SSE=sum(sum((TARGET-OUTPUTVALID').^2));
SSECOL=[SSECOL SSE];
end
[P,Q]=min(SSECOL);
TARGET=[];
for i=1:1:length(INPUTTEST)
    temp=[];
    for j=1:1:length(INPUTTRAIN)
temp=[temp gbf(INPUTTEST(i,:),INPUTTRAIN(j,:),Q(1))];
    end
    TARGET=[TARGET (temp*inv(C)*OUTPUTTRAIN)'];
end
figure(1)
subplot(3,1,1)
plot(TARGET(1,:))
hold on
plot(OUTPUTTEST(:,1),'r')
```

```
subplot(3,1,2)
plot(TARGET(2,:))
hold on
plot(OUTPUTTEST(:,2),'r')
subplot(3,1,3)
plot(TARGET(3,:))
hold on
plot(OUTPUTTEST(:,3),'r')
```

Chapter 4
Probabilistic Supervised Classifier and Unsupervised Clustering

4.1 Probabilistic Discriminative Model: Soft-Max Model (Logistic Regression) for Posterior Density Function

Let $\mathbf{x_{ij}}$ be the ith vector belonging to the jth class. Let the total number of classes be r. The probability that the vector $\mathbf{x_i}$ belonging to the jth class is computed as follows:

$$\mathbf{y_{ji}} = y_j(\mathbf{x_i}) = \frac{e^{-(w_i^T \phi(\mathbf{x})_j)\phi}}{\sum_{m=1}^{m=r} e^{-(w_m^T(\mathbf{x}_j))T}} \tag{4.1}$$

where ϕ_j is the basis vector presented as $\left[\phi_0(\mathbf{x}_j)\ \phi_1(\mathbf{x}_j) \ldots\ \phi_{M-1}(\mathbf{x}_j)\right]^T$.

Consider the training set $\mathbf{x_1}, \mathbf{x_2}, \ldots, \mathbf{x_{100}}$ and let the individual vector $\mathbf{x_i}$ belongs to any one of the $r = 4$ classes. Construct the matrix T with size 100×4, such that the first 25 rows are filled up with [1 0 0 0], the next 25 rows are filled up with [0 1 0 0], the next 25 rows are filled up with [0 0 1 0], and the last 25 rows are filled up with [0 0 0 1]. The matrix T is the target matrix, with $(i, j)^{th}$ element is the target element corresponding to the variable y_{ji}.

4.1.1 Entropy and Cross-Entropy

Let X be the random variable, which have the outcomes X_1, X_2, X_3, and X_4 with probabilities p_1, p_2, p_3, and p_4, respectively. The average information per symbol associated with the random variable X is given as follows:

$$-\sum_{i=1}^{i=4} p_i log(p_i) \tag{4.2}$$

© Springer Nature Switzerland AG 2020

E. S. Gopi, *Pattern Recognition and Computational Intelligence Techniques Using Matlab*, Transactions on Computational Science and Computational Intelligence,
https://doi.org/10.1007/978-3-030-22273-4_4

Let the probability associated with the outcomes of the random variable X differs from the original values and is identified as q_i's. To check how close the identified probabilities are closer with the true probabilities, the cross-entropy is used as given as follows:

$$-\sum_{i=1}^{i=4} p_i log(q_i) \tag{4.3}$$

The cross-entropy is always higher than the actual entropy associated with the typical random variable X (Fig. 4.1). Hence this is used as the objective function.

crossentropy.m

```
%Importance of cross-entropy
temp=rand(1,5);
p=temp/sum(temp);
Entropy=sum(-p.*log(p));
COL=[];
for i=1:1:100
temp=rand(1,5);
q=temp/sum(temp);
COL=[COL sum(-p.*log(q))]
```

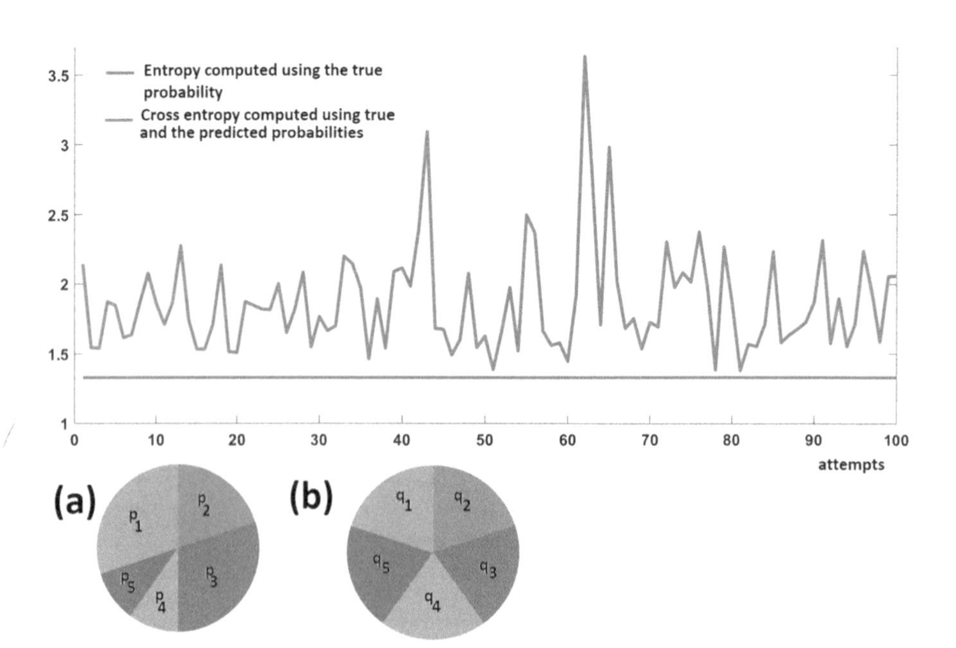

Fig. 4.1 Entropy versus cross-entropy. (**a**) Typical true probabilities of the partition set. (**b**) Typical predicted probabilities of the partition set

```
end
figure
plot(Entropy*ones(1,100))
hold on
plot(COL,'r')
```

The objective function J is obtained by taking negative logarithm of the likelihood function. $\pi_{i=1}^{i=100}\pi_{j=1}^{j=4}y_{ij}^{t_{ij}}$ as follows:

$$J = -\sum_{i=1}^{i=100}\sum_{j=1}^{j=4} t_{ij}log(y_{ij}) \tag{4.4}$$

The requirement is to optimize $\mathbf{w_k}$ in (4.5) for $k = 1\cdots 4$ by maximizing (4.4). Construct the variable $t_{ij} = 1$, if i^{th} vector belongs to the jth class.

1. Case 1: $k = i$

$$\mathbf{y_{ij}} = y_i(\mathbf{x_j}) = \frac{e^{-(w_i^T\phi(\mathbf{x_j}))}}{\sum_{m=1}^{m=r} e^{-(w_m^T\phi(\mathbf{x_j}))}} \tag{4.5}$$

$$-\frac{\left(\sum_{m=1}^{m=r} e^{-(w_m^T T(\mathbf{x_j}))}\right) e^{-(w_i^T T(\mathbf{x_j}))} T(\mathbf{x_j}) - e^{-(w_i^T T(\mathbf{x_j}))} e^{-(w_k^T T(\mathbf{x_j}))} T(\mathbf{x_j})}{\left(\sum_{m=1}^{m=r} e^{-(w_m^T T(\mathbf{x_j}))}\right)^2} \tag{4.6}$$

$$= -y_{ij}\phi(\mathbf{x_j}) - y_{ij}y_{kj}\phi(\mathbf{x_j}) \tag{4.7}$$

$$= -y_{ij}\phi(\mathbf{x_j})(1 - y_{kj}) \tag{4.8}$$

2. Case 2: $k \neq i$, we get the following:

$$-\ y_{ij}\phi(\mathbf{x_j})(0 - y_{kj}) \tag{4.9}$$

Combining (4.8) and (4.9), we get the following:

$$-\ y_{ij}\phi(\mathbf{x_j})(I_{ki} - y_{kj}) \tag{4.10}$$

where I_{ki} is equal to 1 if $k = i$, 0, otherwise. Differentiating (4.4) with respect to $\mathbf{w_k}$, we get the following:

$$\sum_{j=1}^{j=100}\sum_{i=1}^{i=4} \frac{t_{ij}}{y_{ij}} y_{ij}\phi(\mathbf{x_j})(I_{ki} - y_{kj})$$

$$= \sum_{j=1}^{j=100} (t_{1j}\phi(\mathbf{x_j})y_{kj} + t_{2j}\phi(\mathbf{x_j})y_{kj} + t_{3j}$$

$$\times\phi(\mathbf{x_j})y_{kj} + t_{4j}\phi(\mathbf{x_j})y_{kj} + T(\mathbf{x_j})y_{kj} - t_{kj}T(\mathbf{x_j}))$$

$$= \sum_{j=1}^{j=100} \phi(\mathbf{x_j})(y_{kj} - t_{kj}) \tag{4.11}$$

Thus the optimal value of the vector $\mathbf{w_k}$ is obtained using the steepest descent–based iterative technique as follows:

$$\mathbf{w_k}(t+1) = \mathbf{w_k}(t+1) - \eta \sum_{j=1}^{j=100} \phi(\mathbf{x_j})(y_{kj} - t_{kj}) \tag{4.12}$$

The steps involved in logistic regression (for four class problem) are as follows:

1. Initialize the weight vectors $\mathbf{w_k}$ for $k = 1 \cdots 4$.
2. Compute y_{kj} for $k = 1 \cdots 4$ and $j = 1 \cdots 100$.
3. Update the weight vectors using $\mathbf{w_k}(t+1) = \mathbf{w_k}(t+1) - \eta \sum_{j=1}^{j=100} \phi(\mathbf{x_j})(y_{kj} - t_{kj})$.
4. Repeat (2) and (3) for finite number of iterations to obtain the optimal weight vectors $\mathbf{w_k}$ for $k = 1 \cdots 4$.

logisticregression.m

```
%Probabilistic approach to classification - Logistic
approach
m1=[1 0 0]';
m2=[0 1 0]';
m3=[0 0 1]';
m1=rand(3,1)+m1;
m2=rand(3,1)+m2;
m3=rand(3,1)+m3;
N1=1000; N2=1000; N3=1000;
POSTRAIN=[];
POSTEST=[];
POSVAL=[];
temp1=rand(length(m1),1);
C1=temp1*temp1'*0.2;
temp2=rand(length(m2),1);
C2=temp2*temp2'*0.2;
temp3=rand(length(m3),1);
C3=temp3*temp3'*0.2;
[X1,C1est]=genrandn(m1,C1,N1);
[X2,C2est]=genrandn(m2,C2,N2);
[X3,C3est]=genrandn(m3,C3,N3);
DATA=[X1 X2 X3];
[V,M]=kmeans(DATA',3);
```

```
sigma=1:0.5:5;
sigma=1;
for trial=1:1:length(sigma)
B1=[];
B2=[];
B3=[];
for i=1:1:size(DATA,2)
B1=[B1 gbf(DATA(:,i)',M(1,:),sigma(trial))];
B2=[B2 gbf(DATA(:,i)',M(2,:),sigma(trial))];
B3=[B3 gbf(DATA(:,i)',M(3,:),sigma(trial))];
end
%Formulation of matrix
MAT=[B1' B2' B3'];
TRAIN=[MAT(1:1:400,:); MAT(1001:1:1400,:); MAT(2001:
1:2400,:)];
TRAININDEX=zeros(1200,3);
TRAININDEX(1:400,1)=1;
TRAININDEX(401:800,2)=1;
TRAININDEX(801:1200,3)=1;
VAL=[MAT(401:1:700,:); MAT(1401:1:1700,:); MAT(2401:
1:2700,:)];
VALINDEX=zeros(900,3);
VALINDEX(1:300,1)=1;
VALINDEX(301:600,2)=1;
VALINDEX(601:900,3)=1;
%Initializing the weights
W=rand(3,3);
%Soft max computation for the individual classses
P1=exp(TRAIN*W(:,1))./(exp(TRAIN*W(:,1))+exp(TRAIN*
W(:,2))+exp(TRAIN*W(:,3)));
P2=exp(TRAIN*W(:,2))./(exp(TRAIN*W(:,1))+exp(TRAIN*
W(:,2))+exp(TRAIN*W(:,3)));
P3=exp(TRAIN*W(:,3))./(exp(TRAIN*W(:,1))+exp(TRAIN*
W(:,2))+exp(TRAIN*W(:,3)));
%Error function computation
P=[P1 P2 P3];
error=-1*sum(sum(TRAININDEX.*log(P)));
ERROR=[];
for iteration=1:1:1000
W(:,1)=W(:,1)-0.01*TRAIN'*(P(:,1)-TRAININDEX(:,1));
W(:,2)=W(:,2)-0.01*TRAIN'*(P(:,2)-TRAININDEX(:,2));
W(:,3)=W(:,3)-0.01*TRAIN'*(P(:,3)-TRAININDEX(:,3));
%Soft max computation for the individual classses
P1=exp(TRAIN*W(:,1))./(exp(TRAIN*W(:,1))+exp(TRAIN*W
(:,2))+exp(TRAIN*W(:,3)));
```

```
P2=exp(TRAIN*W(:,2))./(exp(TRAIN*W(:,1))+exp(TRAIN*W
(:,2))+exp(TRAIN*W(:,3)));
P3=exp(TRAIN*W(:,3))./(exp(TRAIN*W(:,1))+exp(TRAIN*W
(:,2))+exp(TRAIN*W(:,3)));
%Error function computation
P=[P1 P2 P3];
error=-1*sum(sum(TRAININDEX.*log(P)));
ERROR=[ERROR error];
end
[V,I]=max(P');
REF=[ones(1,400) ones(1,400)*2 ones(1,400)*3];
[R,L]=find((I-REF)==0);
POSTRAIN=[POSTRAIN (length(L)/length(I))*100];
P1=exp(VAL*W(:,1))./(exp(VAL*W(:,1))+exp(VAL*W(:,2))
+exp(VAL*W(:,3)));
P2=exp(VAL*W(:,2))./(exp(VAL*W(:,1))+exp(VAL*W(:,2))
+exp(VAL*W(:,3)));
P3=exp(VAL*W(:,3))./(exp(VAL*W(:,1))+exp(VAL*W(:,2))
+exp(VAL*W(:,3)));
P=[P1 P2 P3];
[V,I]=max(P');
REF=[ones(1,300) ones(1,300)*2 ones(1,300)*3];
[R,L]=find((I-REF)==0);
POSVAL=[POSVAL (length(L)/length(I))*100];
end
[V,I]=max(POSVAL);
%Using testing data
B1=[];
B2=[];
B3=[];
for i=1:1:size(DATA,2)
B1=[B1 gbf(DATA(:,i)',M(1,:),sigma(I))];
B2=[B2 gbf(DATA(:,i)',M(2,:),sigma(I))];
B3=[B3 gbf(DATA(:,i)',M(3,:),sigma(I))];
end
%Formulation of matrix
MAT=[B1' B2' B3'];
TEST=[MAT(701:1:1000,:); MAT(1701:1:2000,:);
MAT(2701:1:3000,:)];
TESTINDEX=zeros(1200,3);
TESTINDEX(1:300,1)=1;
TESTINDEX(301:600,2)=1;
TESTINDEX(601:900,3)=1;

P1=exp(TEST*W(:,1))./(exp(TEST*W(:,1))+exp(TEST*W
```

```
(:,2))+exp(TEST*W(:,3)));
P2=exp(TEST*W(:,2))./(exp(TEST*W(:,1))+exp(TEST*W
(:,2))+exp(TEST*W(:,3)));
P3=exp(TEST*W(:,3))./(exp(TEST*W(:,1))+exp(TEST*W
(:,2))+exp(TEST*W(:,3)));
P=[P1 P2 P3];
[V,I]=max(P');
REF=[ones(1,300) ones(1,300)*2 ones(1,300)*3];
[R,L]=find((I-REF)==0);
POSTEST=[length(L)/length(I)]*100;
```

4.1.2 Two-Class Logistic Regression as the Iterative Recursive Least Square Method

Consider the two-class problem. The function $y(\mathbf{x_n})$ is formulated as follows:

$$t_n = y(\mathbf{x_n}) = \frac{1}{1 + e^{-\mathbf{w^T}\boldsymbol{\phi}(\mathbf{x_n})}} \tag{4.13}$$

The target corresponding to class 1 is assigned as 1, and for the class 2, it is assigned as 0. The optimal **w** vector is obtained by maximizing the objective function J as follows:

$$J = -\sum_{n=1}^{n=N} t_n log(y_n) + (1 - t_n) log(1 - y_n) \tag{4.14}$$

The gradient of J is obtained as

$$\sum_{n=1}^{n=N} (y_n - t_n)\phi(\mathbf{x_n}) = \phi^T(y - t)$$

4.1.2.1 Newton–Raphson Method

Given the particular initialization value **x**, the function $f(\mathbf{x})$ is approximated using the first three terms of the Taylor series. In this technique, the next best choice for x (i.e., x_{next}) is obtained by minimizing the approximated function $g(\mathbf{x})$ of $f(\mathbf{x})$. This is repeated for finite number of iterations to obtain the solution for the structured problem. To obtain the minimum of the approximate function, $g(\mathbf{x})$, we differentiate with respect to **x** and equate to zero.

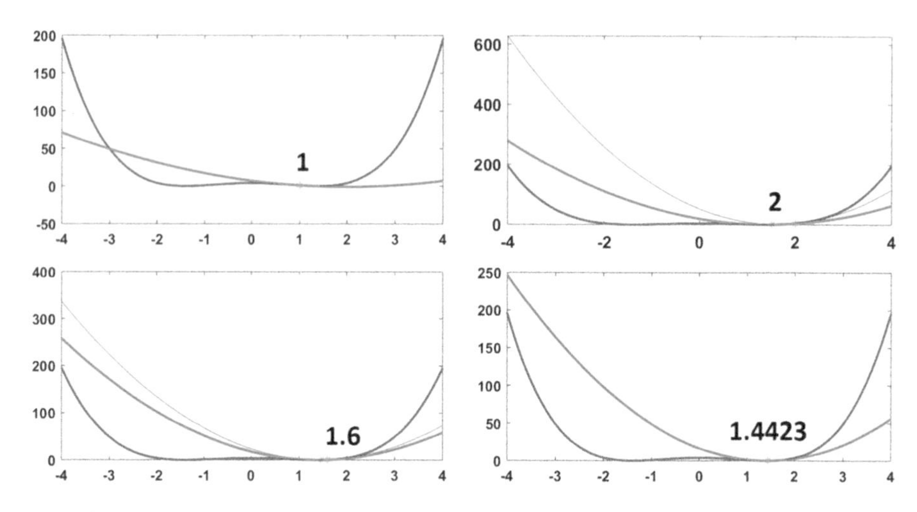

Fig. 4.2 Demonstration of Newton–Raphson method

$$g(\mathbf{x}) = f(\mathbf{x_o}) + \frac{(\mathbf{x}-\mathbf{x_o})^{\mathbf{T}}}{\mathbf{1}!}\nabla\mathbf{f} + \frac{(\mathbf{x}-\mathbf{x_o})^{T}\mathbf{H}(\mathbf{x}-\mathbf{x_o})}{2!} \tag{4.15}$$

This is known as Newton–Raphson method (Fig. 4.2) and is illustrated as follows.

newtonraphson.m

```
\subsubsection{Illustration}
%Demonstration of Newton--Raphson algorithm
x0=1;
COL=[];
for k=1:1:4
COL=[COL x0];
y=[];
z=[];
for x=-4:0.1:4;
y=[y f(x)];
z=[z f(x0)+(x-x0)*f1(x0)+(f2(x0)*((x-x0)^2)/2)];
end
x=-4:0.1:4;
figure(1)
subplot(2,2,k)
plot(x,y,'b')
hold on
plot(x,z,'r')
plot(x0,f(x0),'*')
xnext=x0-(f1(x0)/f2(x0));
x0=xnext;
```

```
end
figure
stem(COL)
```

$$\nabla \mathbf{f} + (\mathbf{x} - \mathbf{x_0})^T \mathbf{H} = 0 \Rightarrow \mathbf{x} = \mathbf{x_0} - \mathbf{H}^{-1} \nabla \mathbf{f} \tag{4.16}$$

To use Newton–Raphson method, the Hessian matrix for the objective function (4.14) is obtained by differentiating (4.14) with respect to **w**. We get the following:

$$\sum_{n=1}^{n=N} y_n(1 - y_n)\phi(x_n)^T \phi(x_n$$

$$= \Phi^T R \Phi$$

where the diagonal elements of the matrix **R** is represented as $y_n(1 - y_n)$. The final equation using the Newton–Raphson method is given as follows:

$$\mathbf{w_{next}} = \mathbf{w_{cur}} - (\Phi^T R \Phi)^{-1} (\Phi)^T (y - t) \tag{4.17}$$

$$\Rightarrow \mathbf{w_{next}} = (\Phi^T R \Phi)^{-1} ((\Phi^T R \Phi)\mathbf{w_{cur}} - (\Phi)^T (y - t)) \tag{4.18}$$

$$\Rightarrow \mathbf{w_{next}} = (\Phi^T R \Phi)^{-1} (\Phi^T R (\mathbf{w_{cur}} - R^{-1} (y - t)) \tag{4.19}$$

$$\Rightarrow \mathbf{w_{next}} = (\Phi^T R \Phi)^{-1} \Phi^T R z \tag{4.20}$$

4.1.2.2 Observing (4.20) as the Recursive Weighted Least Square Method

Consider the problem of least square that minimizes $(\mathbf{t} - \boldsymbol{\phi}\mathbf{w})^T (\mathbf{t} - \boldsymbol{\phi}\mathbf{w_{cur}})$. The optimal solution to the unknown vector **w** is obtained as follows:

$$\mathbf{w} = (\boldsymbol{\phi}^T \boldsymbol{\phi})^{-1} (\boldsymbol{\phi}^T)\mathbf{t} \tag{4.21}$$

Consider another problem of least square that minimizes

$$\left(\sqrt{\mathbf{R}}\mathbf{t} - \sqrt{\mathbf{R}}\boldsymbol{\phi}\mathbf{w}\right)^T \left(\sqrt{\mathbf{R}}\mathbf{t} - \sqrt{\mathbf{R}}\boldsymbol{\phi}\mathbf{w}\right) \tag{4.22}$$

$$(\mathbf{t} - \boldsymbol{\phi}\mathbf{w})^T \mathbf{R} (\mathbf{t} - \boldsymbol{\phi}\mathbf{w}) \tag{4.23}$$

The optimal solution to the unknown user **w** is obtained as follows:

$$\mathbf{w} = (\boldsymbol{\phi}^T \mathbf{R} \boldsymbol{\phi})^{-1} (\boldsymbol{\phi}^T \mathbf{R} \mathbf{t}) \tag{4.24}$$

In the least square technique, we assume that the variance of the noises added in each observations is identical. In the case of weighted least square technique, weights are chosen as the estimated inverse of the noise variance involved in every observation. Comparing (4.20) and (4.24), we understand that the Newton method applied to two-class logistic regression acts like the weighted least square solution. In every iteration, the target vector **z** is modified according to the latest weight vector. Hence this is known as Iterative Recursive Least Square (IRLS) method.

4.2 Newton–Raphson Method for Multi-Class Problem

Differentiating (4.11) with respect to $\mathbf{w_k}$, we get the following:

$$\sum_{j=1}^{j=100} y_{kj}\phi(\mathbf{x_j})\phi(\mathbf{x_j})^T(1-y_{kj}) \tag{4.25}$$

Thus to update $\mathbf{w_k}$, the following equation is used.

$$\mathbf{w_k}(t+1)=\mathbf{w_k}(t+1)-\eta\left(\sum_{j=1}^{j=100} y_{kj}\phi(\mathbf{x_j})\phi(\mathbf{x_j})^T(1-y_{kj})\right)^{-1}\sum_{j=1}^{j=100} T(\mathbf{x_j})(y_{kj}-t_{kj}) \tag{4.26}$$

Using gradient and Newton–Raphson methods, logistic regression is illustrated in Figs. 4.3–4.7.

logisticregression-newtonraphson.m

```
%Probabilistic approach to classification - Logistic
approach
%Using Newton Raphson method
m1=[1 0 0]';
m2=[0 1 0]';
m3=[0 0 1]';
m1=rand(3,1)+m1;
m2=rand(3,1)+m2;
m3=rand(3,1)+m3;
N1=1000; N2=1000; N3=1000;
```

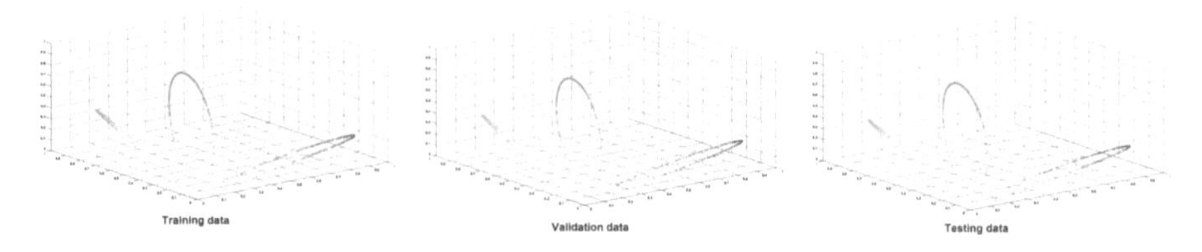

Fig. 4.3 Data used to demonstrate multi-class logistic regression using gradient method

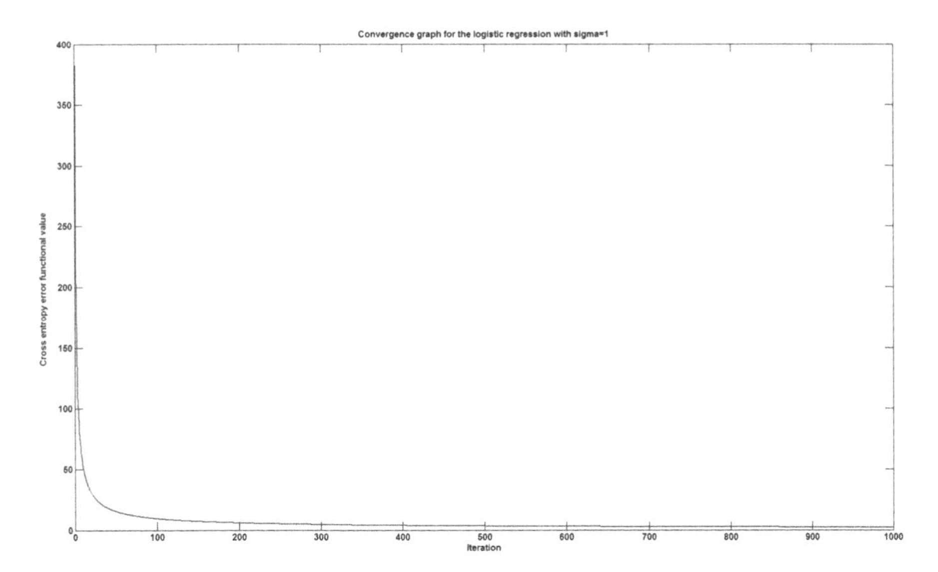

Fig. 4.4 Convergence graph of multi-class logistic regression

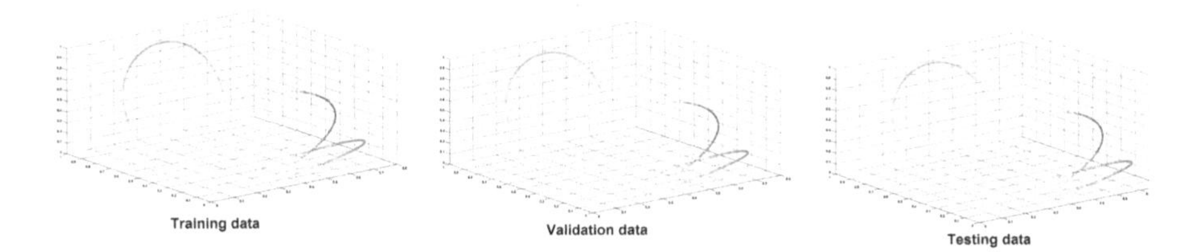

Fig. 4.5 Data used for demonstrating Newton–Raphson method for multi-class logistic regression

```
POSTRAIN=[];
POSTEST=[];
POSVAL=[];
POSTRAINH=[];
POSTESTH=[];
POSVALH=[];
ERROR=[];
ERRORH=[];
temp1=rand(length(m1),1);
C1=temp1*temp1'*0.2;
temp2=rand(length(m2),1);
C2=temp2*temp2'*0.2;
temp3=rand(length(m3),1);
C3=temp3*temp3'*0.2;
[X1,C1est]=genrandn(m1,C1,N1);
[X2,C2est]=genrandn(m2,C2,N2);
```

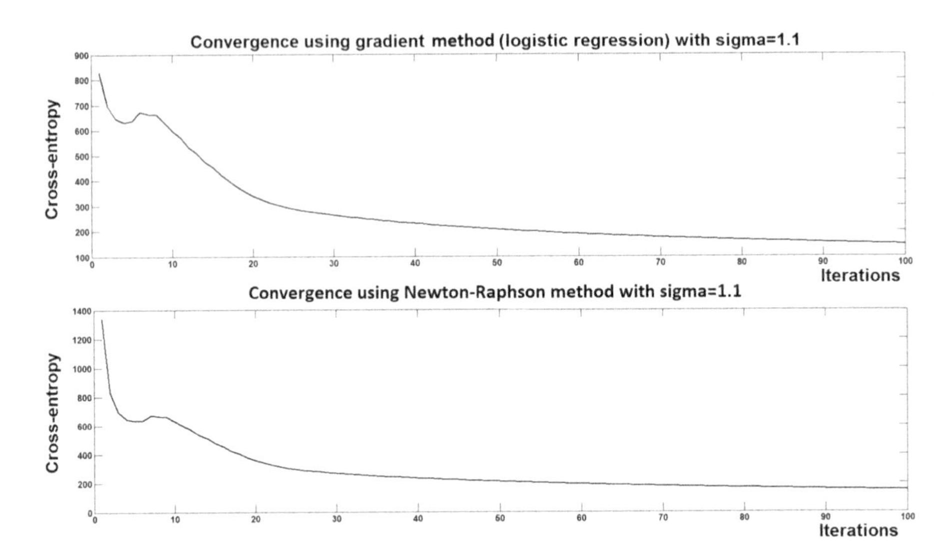

Fig. 4.6 Convergence of logistic regression-steepest descent algorithm versus Newton–Raphson method

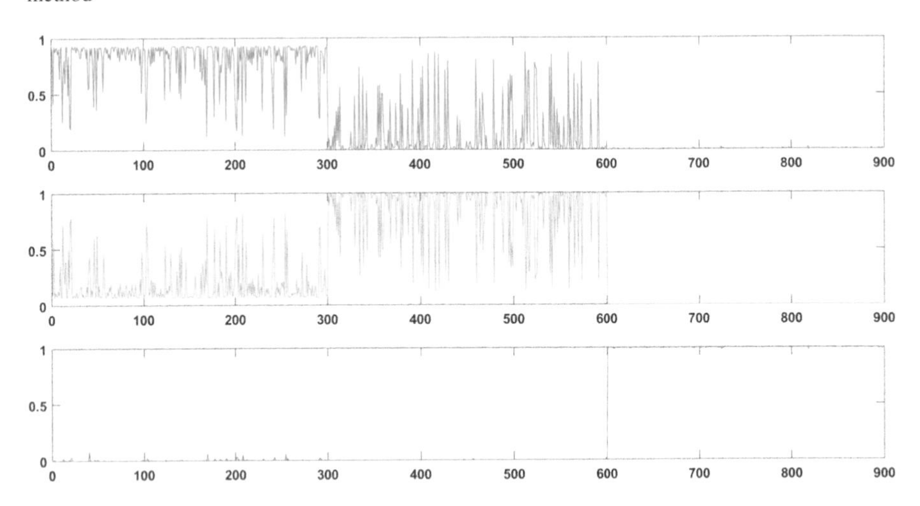

Fig. 4.7 Illustration of the actual probabilities obtained using logistic regression for the three-class test data

```
[X3,C3est]=genrandn(m3,C3,N3);
DATA=[X1 X2 X3];
[V,M]=kmeans(DATA',3);
sigma=1:0.5:5;
sigma=1;
for trial=1:1:length(sigma)
B1=[];
```

```
B2=[];
B3=[];
for i=1:1:size(DATA,2)
B1=[B1 gbf(DATA(:,i)',M(1,:),sigma(trial))];
B2=[B2 gbf(DATA(:,i)',M(2,:),sigma(trial))];
B3=[B3 gbf(DATA(:,i)',M(3,:),sigma(trial))];
end
%Formulation of matrix
MAT=[B1' B2' B3'];
TRAIN=[MAT(1:1:400,:); MAT(1001:1:1400,:); MAT(2001:
1:2400,:)];
TRAININDEX=zeros(1200,3);
TRAININDEX(1:400,1)=1;
TRAININDEX(401:800,2)=1;
TRAININDEX(801:1200,3)=1;
VAL=[MAT(401:1:700,:); MAT(1401:1:1700,:); MAT(2401:
1:2700,:)];
VALINDEX=zeros(900,3);
VALINDEX(1:300,1)=1;
VALINDEX(301:600,2)=1;
VALINDEX(601:900,3)=1;
%Initializing the weights
W=rand(3,3);
WH=W;
%Soft max computation for the individual classses
P1=exp(TRAIN*W(:,1))./(exp(TRAIN*W(:,1))+exp(TRAIN*W
(:,2))+exp(TRAIN*W(:,3)));
P2=exp(TRAIN*W(:,2))./(exp(TRAIN*W(:,1))+exp(TRAIN*W
(:,2))+exp(TRAIN*W(:,3)));
P3=exp(TRAIN*W(:,3))./(exp(TRAIN*W(:,1))+exp(TRAIN*W
(:,2))+exp(TRAIN*W(:,3)));
%Error function computation
P=[P1 P2 P3];
error=-1*sum(sum(TRAININDEX.*log(P)));
ERROR=[];
for iteration=1:1:100
PH=P;
I=diag([1 1 1]);
temp=0;
H=[];
for r=1:1:3
    for s=1:1:3
        temp=0;
    for n=1:1:1200
temp=temp+TRAIN(n,:)'*TRAIN(n,:)*PH(n,r)*(I(r,s)-PH
```

```
(n,s));
    end
HESSIAN{r,s}=-temp;
    end
end
WH(:,1)=WH(:,1)-0.01*pinv(HESSIAN{1,1})*TRAIN'*(P(:,1)
-TRAININDEX(:,1));
WH(:,2)=WH(:,2)-0.01*pinv(HESSIAN{2,2})*TRAIN'*(P(:,2)
-TRAININDEX(:,2));
WH(:,3)=WH(:,3)-0.01*pinv(HESSIAN{3,3})*TRAIN'*(P(:,3)
-TRAININDEX(:,3));
%Soft max computation for the individual classses
PH1=exp(TRAIN*WH(:,1))./(exp(TRAIN*WH(:,1))+...
exp(TRAIN*WH(:,2))+exp(TRAIN*WH(:,3)));
PH2=exp(TRAIN*WH(:,2))./(exp(TRAIN*WH(:,1))+...
exp(TRAIN*WH(:,2))+exp(TRAIN*WH(:,3)));
PH3=exp(TRAIN*WH(:,3))./(exp(TRAIN*WH(:,1))+...
exp(TRAIN*WH(:,2))+exp(TRAIN*WH(:,3)));
%Error function computation
PH=[PH1 PH2 PH3];
error=-1*sum(sum(TRAININDEX.*log(P)));
ERRORH=[ERRORH error];
W(:,1)=W(:,1)-0.01*TRAIN'*(P(:,1)-TRAININDEX(:,1));
W(:,2)=W(:,2)-0.01*TRAIN'*(P(:,2)-TRAININDEX(:,2));
W(:,3)=W(:,3)-0.01*TRAIN'*(P(:,3)-TRAININDEX(:,3));
%Soft max computation for the individual classses
P1=exp(TRAIN*W(:,1))./(exp(TRAIN*W(:,1))+...
exp(TRAIN*W(:,2))+exp(TRAIN*W(:,3)));
P2=exp(TRAIN*W(:,2))./(exp(TRAIN*W(:,1))+...
exp(TRAIN*W(:,2))+exp(TRAIN*W(:,3)));
P3=exp(TRAIN*W(:,3))./(exp(TRAIN*W(:,1))+...
exp(TRAIN*W(:,2))+exp(TRAIN*W(:,3)));
%Error function computation
P=[P1 P2 P3];
error=-1*sum(sum(TRAININDEX.*log(P)));
ERROR=[ERROR error];
end
figure
plot(ERROR,'r')
hold on
plot(ERROR,'b')
[V,I]=max(PH');
REF=[ones(1,400) ones(1,400)*2 ones(1,400)*3];
[R,L]=find((I-REF)==0);
POSTRAINH=[POSTRAINH (length(L)/length(I))*100];
```

```
PH1=exp(VAL*WH(:,1))./(exp(VAL*WH(:,1))+...
exp(VAL*WH(:,2))+exp(VAL*WH(:,3)));
PH2=exp(VAL*WH(:,2))./(exp(VAL*WH(:,1))+...
exp(VAL*WH(:,2))+exp(VAL*WH(:,3)));
PH3=exp(VAL*WH(:,3))./(exp(VAL*WH(:,1))+...
exp(VAL*WH(:,2))+exp(VAL*WH(:,3)));
PH=[PH1 PH2 PH3];
[V,I]=max(PH');
REF=[ones(1,300) ones(1,300)*2 ones(1,300)*3];
[R,L]=find((I-REF)==0);
POSVALH=[POSVALH (length(L)/length(I))*100];
%%%%%%%%%%%%%%%%%%%%%%%%%%%%%%%%%%%%%%%%%%%%%%%%%%%%
%%%%%%%%%%%%%%%%%%%%%%
[V,I]=max(P');
REF=[ones(1,400) ones(1,400)*2 ones(1,400)*3];
[R,L]=find((I-REF)==0);
POSTRAIN=[POSTRAIN (length(L)/length(I))*100];
P1=exp(VAL*W(:,1))./(exp(VAL*W(:,1))+...
exp(VAL*W(:,2))+exp(VAL*W(:,3)));
P2=exp(VAL*W(:,2))./(exp(VAL*W(:,1))+...
exp(VAL*W(:,2))+exp(VAL*W(:,3)));
P3=exp(VAL*W(:,3))./(exp(VAL*W(:,1))+...
exp(VAL*W(:,2))+exp(VAL*W(:,3)));
P=[P1 P2 P3];
[V,I]=max(P');
REF=[ones(1,300) ones(1,300)*2 ones(1,300)*3];
[R,L]=find((I-REF)==0);
POSVAL=[POSVAL (length(L)/length(I))*100];
end
[V,I]=max(POSVAL);
%Using testing data
B1=[];
B2=[];
B3=[];
for i=1:1:size(DATA,2)
B1=[B1 gbf(DATA(:,i)',M(1,:),sigma(I))];
B2=[B2 gbf(DATA(:,i)',M(2,:),sigma(I))];
B3=[B3 gbf(DATA(:,i)',M(3,:),sigma(I))];
end
%Formulation of matrix
MAT=[B1' B2' B3'];
TEST=[MAT(701:1:1000,:); MAT(1701:1:2000,:);
MAT(2701:1:3000,:)];
TESTINDEX=zeros(1200,3);
TESTINDEX(1:300,1)=1;
```

```
TESTINDEX(301:600,2)=1;
TESTINDEX(601:900,3)=1;

P1=exp(TEST*W(:,1))./(exp(TEST*W(:,1))+exp(TEST*W
(:,2))+exp(TEST*W(:,3)));
P2=exp(TEST*W(:,2))./(exp(TEST*W(:,1))+exp(TEST*W
(:,2))+exp(TEST*W(:,3)));
P3=exp(TEST*W(:,3))./(exp(TEST*W(:,1))+exp(TEST*W
(:,2))+exp(TEST*W(:,3)));
P=[P1 P2 P3];
[V,I]=max(P');
REF=[ones(1,300) ones(1,300)*2 ones(1,300)*3];
[R,L]=find((I-REF)==0);
POSTEST=[length(L)/length(I)]*100;
%%%%%%%%%%%%%%%%%%%%%%%%%%%%%%%%%%%%%%%%%%
%%%%%%%%%%%%%%%%%%%%%%%%%%%%%%%%
PH1=exp(TEST*WH(:,1))./(exp(TEST*WH(:,1))+exp(TEST*WH
(:,2))+exp(TEST*WH(:,3)));
PH2=exp(TEST*WH(:,2))./(exp(TEST*WH(:,1))+exp(TEST*WH
(:,2))+exp(TEST*WH(:,3)));
PH3=exp(TEST*WH(:,3))./(exp(TEST*WH(:,1))+exp(TEST*WH
(:,2))+exp(TEST*WH(:,3)));
PH=[PH1 PH2 PH3];
[V,I]=max(PH');
REF=[ones(1,300) ones(1,300)*2 ones(1,300)*3];
[R,L]=find((I-REF)==0);
POSTESTH=[length(L)/length(I)]*100;
figure
subplot(2,1,1)
subplot(ERROR)
title('Convergence using gradient technique')
subplot(2,1,2)
subplot(ERRORH)
title('Convergence using Newton Raphson technique')
```

4.3 Probabilistic Generative Model: Gaussian Model for Class Conditional Density Function

The prior density function of the class labels is given as $p(c_1)$, $p(c_2)$, ..., $p(c_r)$, where r is the number of classes. The posterior probability of the class labels given the observation vector $\mathbf{x}$ is given as $p(c_k/\mathbf{x})$. This is usually not known. We use the parametric approach to model the likelihood function $p(\mathbf{x}/c_k)$ as class conditional probability density function (ccpdf). This is otherwise known as generating model.

The probability that the arbitrary vector $\mathbf{x}$ being generated from the generating model is given as $p(\mathbf{x}/c_k)$. We model the ccpdf as the multivariate Gaussian density function, which is described by its mean vector $\boldsymbol{\mu}_\mathbf{k}$ and co-variance matrix $\mathbf{C_K}$. The mean vector and the co-variance matrix of the typical ccpdf are estimated. Given the matrix $\mathbf{T}$ with (i, j) element takes the value 1 if i^{th} vector belongs to the jth class, 0, otherwise. The unknown parameters are obtained by maximizing the generating probability of the training data $\mathbf{x_1}, \mathbf{x_2}, ..., \mathbf{x_N}$ being generated from the model. The generating probability is given as follows:

$$\Pi_{i=1}^{i=N} \Pi_{j=1}^{j=r} (p(c_j)p(\mathbf{x_i}/c_j))^{t_{ij}} \tag{4.27}$$

Taking logarithm, we get the following:

$$\sum_{i=1}^{i=N} \sum_{j=1}^{j=r} t_{ij} log(p(c_j)(p(\mathbf{x_i}/c_j))) \tag{4.28}$$

Differentiating above with respect to μ_k and equating to zero, we get the following estimate for mean vector as follows:

$$\mu_k = \frac{\sum_{i=1}^{i=N} x_i t_{ik}}{\sum_{i=1}^{i=N} t_{ik}} \tag{4.29}$$

Similarly,

$$C_k = \frac{\sum_{i=1}^{i=N} (x_i - \mu_i)(x_i - \mu_i)^T t_{ik}}{\sum_{i=1}^{i=N} t_{ik}} \tag{4.30}$$

The prior probability of the class is also estimated as follows:

$$p(c_k) = \frac{\sum_{i=1}^{i=N} t_{ik}}{\sum_{i=1}^{i=N} \sum_{j=1}^{j=r} t_{ij}} \tag{4.31}$$

4.4 Gaussian Mixture Model (GMM): Combinational Model

In Sect. 4.3, we have assumed that each vector is being generated from one Gaussian density function (belonging to the individual class). In this case, the vector is assumed as being generated from the mixtures of multivariate Gaussian density function (belonging to multiple classes), i.e., all the classes are responsible for the vector being generated from the GMM model. Let the arbitrary vector $\mathbf{x}$ being generated from the GMM is represented as follows:

$$p(\mathbf{x}) = \sum_{k=1}^{k=N} \pi_k p(\mathbf{x}/\boldsymbol{\mu}_\mathbf{k}, \mathbf{C}_\mathbf{k}) \tag{4.32}$$

Let the training set be represented as $\mathbf{x_1}, \mathbf{x_2}, ..., \mathbf{x_N}$. The vectors are assumed as generated independently. The likelihood function $p(\mathbf{x}/\boldsymbol{\mu}_\mathbf{k}\sum_\mathbf{k})$ is Gaussian distributed described with the parameters mean vector $\boldsymbol{\mu}_\mathbf{k}$ and co-variance matrix $\sum_\mathbf{k}$. They are estimated by maximizing the log likelihood function of the individual vectors (independent training set) being generated from the model as follows:

$$log(p(\mathbf{x_1 x_2} \cdots \mathbf{x_d})) = log(\Pi_{n=1}^{n=N} \sum_{k=1}^{k=r} \pi_k p(\mathbf{x_n}/\boldsymbol{\mu}_\mathbf{k}\mathbf{C}_\mathbf{k})) \tag{4.33}$$

$$\sum_{n=1}^{n=N} log\left(\sum_{k=1}^{k=r} \pi_k p(\mathbf{x_n}/\boldsymbol{\mu}_\mathbf{k}\mathbf{C}_\mathbf{k})\right) \tag{4.34}$$

Differentiating (4.34) with respect to μ_m and equating to zero, we get the following:

$$\sum_{n=1}^{n=N} \pi_m \frac{p(\mathbf{x_n}/\boldsymbol{\mu}_\mathbf{m}\mathbf{C}_\mathbf{m})\mathbf{C}_\mathbf{m}(\mathbf{x_n} - \boldsymbol{\mu}_\mathbf{m})}{(\sum_{k=1}^{n=r} \pi_k p(\mathbf{x_n}/\boldsymbol{\mu}_\mathbf{k}\mathbf{C}_\mathbf{k}))} = 0 \tag{4.35}$$

$$\Rightarrow \sum_{n=1}^{k=N} \frac{\pi_m p(\mathbf{x_n}/\boldsymbol{\mu}_\mathbf{m}\mathbf{C}_\mathbf{k})(\mathbf{x_n} - \boldsymbol{\mu}_\mathbf{m})}{\left(\sum_{k=1}^{k=N} \pi_k p(\mathbf{x_n}/\boldsymbol{\mu}_\mathbf{k}\mathbf{C}_\mathbf{k})\right)} = 0 \tag{4.36}$$

Introducing the responsibility factor $r_m(\mathbf{x_n})$ as follows:

$$r_m(\mathbf{x_n}) = \frac{\pi_m p(\mathbf{x_n}/\boldsymbol{\mu}_\mathbf{m}\mathbf{C}_\mathbf{k})}{\left(\sum_{k=1}^{k=N} \pi_k p(\mathbf{x_n}/\boldsymbol{\mu}_\mathbf{k}\mathbf{C}_\mathbf{k})\right)} \tag{4.37}$$

The responsibility factor $r_m(\mathbf{x_n})$ is the measure of responsibility that the vector $\mathbf{x_n}$ is generated from the class index m and is represented as $r_m(\mathbf{x_n})$. Rewriting (4.36) as follows:

$$\sum_{n=1}^{n=N} r_m(\mathbf{x_n})(\mathbf{x_n} - \boldsymbol{\mu}_\mathbf{m}) = 0 \tag{4.38}$$

$$\sum_{n=1}^{n=N} r_m(\mathbf{x_n})\mathbf{x_n} = \sum_{n=1}^{n=N} r_m(\mathbf{x_n})\boldsymbol{\mu}_\mathbf{m} \tag{4.39}$$

$$\Rightarrow \boldsymbol{\mu}_\mathbf{m} = \frac{\sum_{n=1}^{n=N} r_m(\mathbf{x_n})\mathbf{x_n}}{\sum_{n=1}^{n=N} r_m(\mathbf{x_n})} \tag{4.40}$$

In the same fashion, the co-variance matrix $\mathbf{C_m}$ is computed as follows:

$$\Rightarrow \mathbf{C_m} = \frac{\sum_{n=1}^{n=N} r_m(\mathbf{x_n})(\mathbf{x_n} - \mu_\mathbf{m})((\mathbf{x_n} - \mu_\mathbf{m}))^\mathbf{T}}{\sum_{n=1}^{n=N} r_m(\mathbf{x_n})} \tag{4.41}$$

Also the prior probability π_m of the class index m is obtained as follows:

$$\pi_m = \frac{r_m(\mathbf{x_n})}{\sum_{n=1}^{n=N}} r_m(\mathbf{x_n}) \tag{4.42}$$

Thus the unknown parameters in constructing the Gaussian Mixture Model using the given training data are obtained as follows (Fig. 4.8):

1. Initialize the unknown parameters.
2. Compute the responsibility factor using (4.37). This is the expectation stage.
3. Estimate the unknown parameters $\boldsymbol{\mu}_\mathbf{m}$ and $\mathbf{C_m}$ and π_m.
4. Using the estimated values for $\boldsymbol{\mu}_\mathbf{m}$ and $\mathbf{C_m}$, estimate $r_m(\mathbf{x_n})$ using (4.37). This is the maximization stage. Repeat (2)–(4) until convergence occurs.

GeneratedatausingGMM.m

```
%Generate data using GMM
C1=[0.9 0.1;0.1 0.9];
[E1,D1]=eig(C1);
M1=[2 2];
C2=[0.7 0.3;0.3 0.7];
[E2,D2]=eig(C2);
M2=[-2 -2];
%The prior probability of the first Mixture and the
second Mixtures are
%0.2 and 0.8 respectively.
%Generation of 10000 data with additive noise from
constructed Model is as
%given below
figure(1)
DATA=[]
INDEX=[];
for l=1:1:2000
r=rand;
```

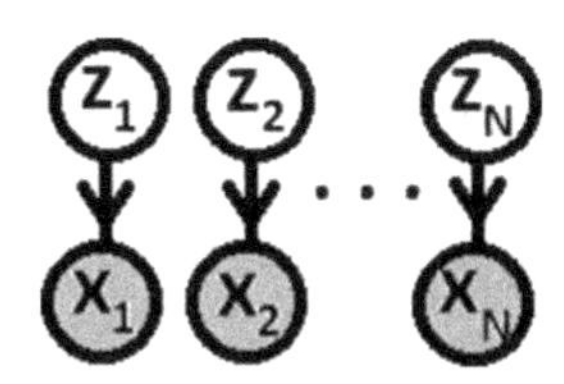

Fig. 4.8 Illustration of GMM model

```
if(r<0.2)
    temp=randn(2,1);
    temp1=E1'*D1.^(1/2)*temp+M1;
    hold on
    plot(temp1(1),temp1(2),'ro');
    %second model
DATA=[DATA temp1];
INDEX=[INDEX 1];
else
    %first model
temp=randn(2,1);
temp2=E2'*D2.^(1/2)*temp+M2;
hold on
plot(temp2(1),temp2(2),'bo');
DATA=[DATA temp2];
INDEX=[INDEX 2];
end
end
%Given data, the mean and the co-variance matrix
of the
%two classes are obtained as follows.
%Initialize mean
[P,Q]=kmeans(DATA',2);
M1=Q(1,:)';
M2=Q(2,:)';
[u,v]=find(P==1);
temp=DATA(:,u);
C1=cov(temp');
C1=C1/2;
p1=length(u)/length(P);
[u,v]=find(P==2);
temp=DATA(:,u);
C2=cov(temp');
p2=length(u)/length(P);
MEAN1COL=M1;
MEAN2COL=M2;
COV1COL{1}=C1;
COV2COL{1}=C2;
PRIOR1COL=p1;
PRIOR2COL=p2;
GP=[];
for iteration=1:1:10
%Expectation stage to estimate the responsibility
factor.
r1=[];
```

```
r2=[];
for i=1:1:length(DATA)
r1=[r1 p1*normalpdf(M1,DATA(:,i),C1)/(p1*normalpdf
(M1,DATA(:,i),C1)+...
    p2*normalpdf(M2,DATA(:,i),C2))];
r2=[r2 p2*normalpdf(M2,DATA(:,i),C2)/(p1*normalpdf
(M1,DATA(:,i),C1)+...
    p2*normalpdf(M2,DATA(:,i),C2))];
end
%Maximization stage
s1=0;
s2=0;
s3=0;
s4=0;
for i=1:1:length(DATA)
s1=s1+DATA(:,i)*r1(i);
s2=s2+DATA(:,i)*r2(i);
end
M1=s1/sum(r1);
M2=s2/sum(r2);
for j=1:1:length(DATA)
s3=s3+(DATA(:,j)-M1')*(DATA(:,j)-M1')'*r1(j);
s4=s4+(DATA(:,j)-M2')*(DATA(:,j)-M2')'*r2(j);
end
C1=s3/(2*sum(r1));
C2=s4/(2*sum(r2));
p1=sum(r1)/length(DATA);
p2=sum(r2)/length(DATA);
% T1=sum((DATA-repmat(M1,1,length(DATA))).^2);
% T2=sum((DATA-repmat(M2,1,length(DATA))).^2);
% [V,I]=min([T1;T2]);
% [u,v]=find(I==1);
% p1=length(v)/length(DATA);
% [u,v]=find(I==2);
% p2=length(v)/length(DATA);
 MEAN1COL=[MEAN1COL M1];
 MEAN2COL=[MEAN2COL M2];
 COV1COL{iteration}=C1;
 COV2COL{iteration}=C2;
 PRIOR1COL=[PRIOR1COL p1];
 PRIOR2COL=[PRIOR2COL p2];
temp=0;
for i=1:1:length(DATA)
    temp=temp+log((p1*normalpdf(M1,DATA(:,i),C1)+...
    p2*normalpdf(M2,DATA(:,i),C2)));
```

```
end
GP=[GP temp];
end
figure
plot(GP)
xlabel('Iteration');
ylabel('Generating probability');
figure
%Given DATA, INDEX estimation is obtained as follows.
l=1;
INDEXEST=[];
for l=1:1:length(DATA)
    p1=(1/2*pi)^(2/2)*(1/(det(C1)^(1/2)))*exp(-(1/2)*
    ((DATA(:,l)-M1')'*(inv(C1))*(DATA(:,l)-M1')));
    p2=(1/2*pi)^(2/2)*(1/(det(C2)^(1/2)))*exp(-(1/2)*
    ((DATA(:,l)-M2')'*(inv(C2))*(DATA(:,l)-M2')));
    if(p1>p2)
        INDEXEST=[INDEXEST 1];
        hold on
        plot(DATA(1,l),DATA(2,l),'r*')
    else
        INDEXEST=[INDEXEST 2];
        hold on
        plot(DATA(1,l),DATA(2,l),'b*')
    end
end
figure
stem(INDEX(1:1:100))
hold on
stem(INDEXEST(1:1:100),'r')
```

normalpdf.m

```
function [res]=normalpdf(x,m,c)
%x is the vector
%m is the mean vector
%c is the co-variance matrix
N=length(x);
temp1=exp(-(1/2)*((x-m)'*inv(abs(c))*(x-m)));
temp2=1/((2*pi)^(N/2));
temp3=1/(det(c)^(1/2));
res=temp1*temp2*temp3;
```

4.4.1 Illustration of GMM

The graphical representation of the Gaussian Mixture Model (GMM) is given in Fig. 4.8. The z_i's are the latent variable corresponding to the ith frame. In this model latent variable takes either [1 0] or [0 1]. If $\mathbf{z_i} = [1\ 0]$, it is interpreted that the ith frame is generated from the state 1. If $\mathbf{z_i} = [0\ 1]$, we interpret that the ith frame is generated from the state 2. The graph also represents that the data generated in different frames are independent. Also being $\mathbf{z_i}$ in the particular state, the arbitrary vector $\mathbf{x}$ being generated from the model is governed by the generating probability. In this case, it is Gaussian density function. The 2D data generated using the model is given in Fig. 4.9. Red and blue represent the data generated using the state 1 and state 2, respectively. The hidden parameters are estimated using EM algorithm by maximizing the generating probability. The convergence of the parameters is illustrated in Fig. 4.10. The increase in the log of the generating probability in each iteration is seen in Fig. 4.11.

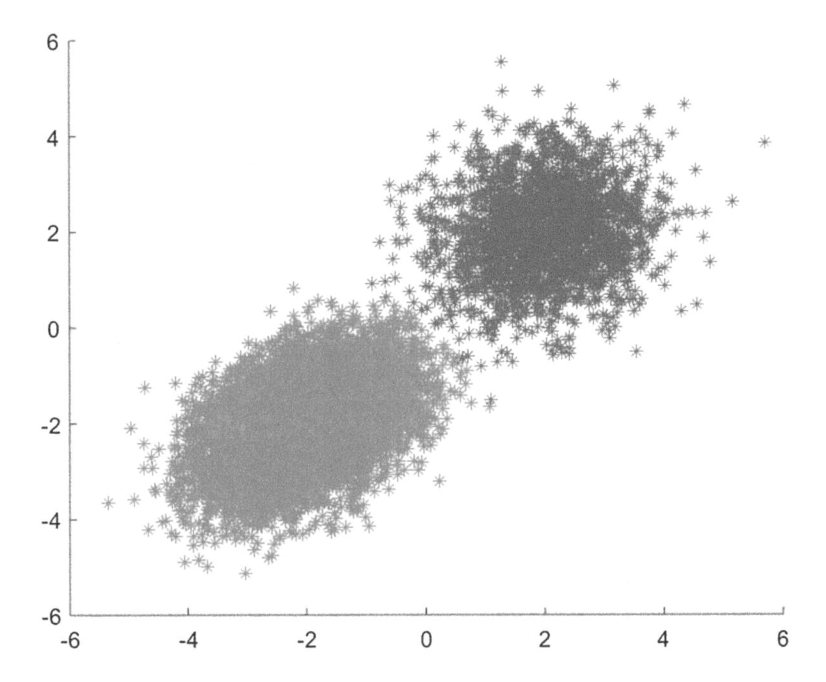

Fig. 4.9 Data used to model GMM

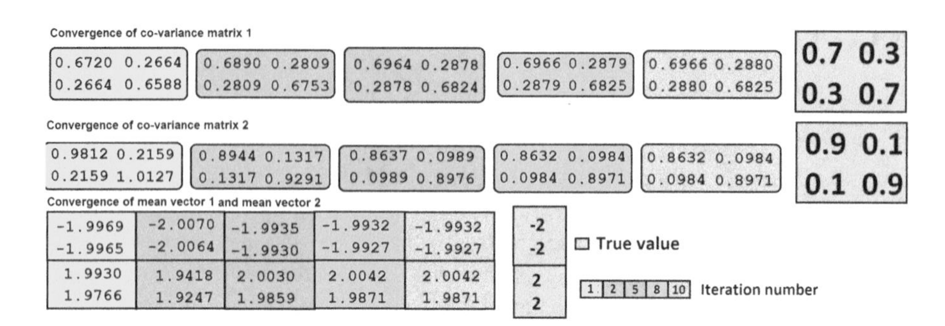

Fig. 4.10 Illustration of convergence obtained using Expectation–Maximization algorithm to GMM

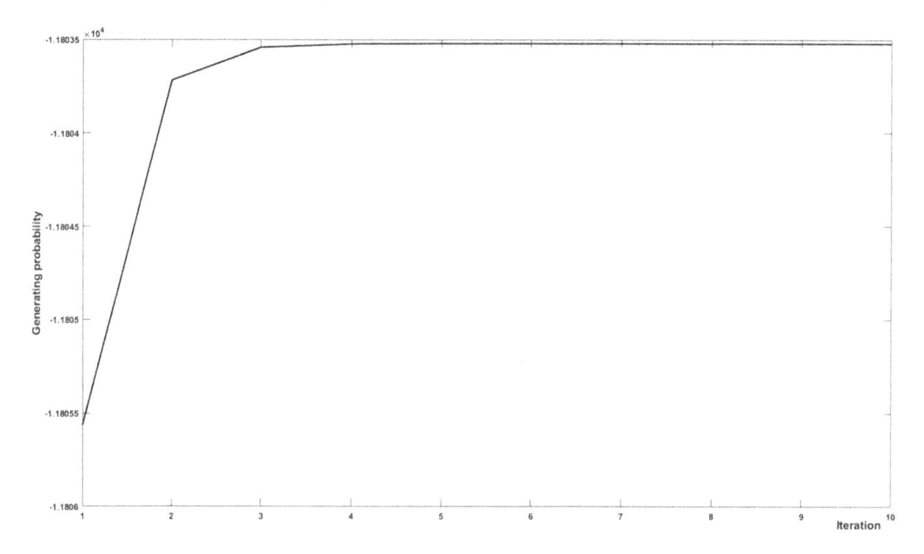

Fig. 4.11 Illustration of maximization of generating probability in modeling GMM using Expectation–Maximization algorithm

4.5 Hidden Markov Model (HMM): Generative Model

The Hidden Markov Model (HMM) is the generative model. Let the sequence of random vectors $\mathbf{x_1}\,\mathbf{x_2}\cdots\mathbf{x_N}$ being generated by the typical model as shown in Fig. 4.12(a).

In the typical example Fig. 4.12b, latent vector $\mathbf{z_1} = z_{11}\ z_{12}\ z_{13}\ z_{14}\ z_{15}$ takes the value [0 1 0 0 0]. This indicates that the latent vector $\mathbf{z_1}$ is in the second state, and it generates $\mathbf{x_1}$. In the same fashion, it is seen that the latent vector $\mathbf{z_2}$ takes the value [0 0 0 1 0], and it indicates that the vector $\mathbf{z_2}$ is in the fourth state. It is seen that there is the state transition from 2 to 4 and being in the state 4, $\mathbf{z_2}$ generates $\mathbf{x_2}$. In the same fashion, the following sequence of actions takes place.

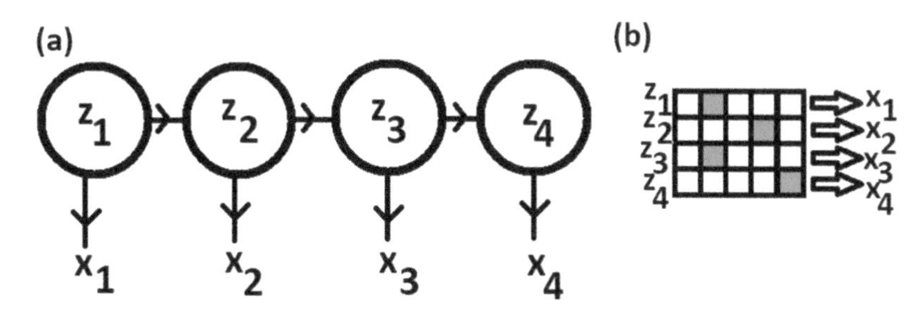

Fig. 4.12 (**a**) Hidden Markov Model. (**b**) Example of HMM model with five states and four time instants

1. There is the state transition from state 4 to 2, i.e., the second element of the vector $\mathbf{z_3}$ takes the value 1 and generates $\mathbf{x_3}$.
2. There is the state transition from state 2 to 5, i.e., the fifth element of the vector $\mathbf{z_4}$ takes the value 1 and generates $\mathbf{x_4}$.

The generating actions are described by the following actions:

1. The probability that the first vector $\mathbf{x_1}$ being generated by the lth state is described by the prior probability π_l and is computed as follows:

$$p(\mathbf{z_1}) = \Pi_{i=1}^{i=4} \pi_i^{z_{1,i}} \tag{4.43}$$

It indicates that $p(\mathbf{z_1})$ takes the value π_l, if the lth element of the vector $\mathbf{z_1}$ is 1.
2. The state transition from the ith state to the jth state is described by the state transition probability A_{ij}.
3. $p(\mathbf{z_n}/\mathbf{z_{n-1}})$ is the transition probability of state changes from $(n-1)^{\text{th}}$ time instant into n^{th} time instant. It takes any of the value among the 16 elements of the matrix $\mathbf{A}$ (4×4 matrix) A_{ij}, which is obtained as follows:

$$p(\mathbf{z_n}/\mathbf{z_{n-1}}) = \Pi_{i=1}^{i=4} \Pi_{j=1}^{i=4} (A_{ij})^{z_{n,i} z_{n-1,j}} \tag{4.44}$$

This indicates that $p(\mathbf{z_n}/\mathbf{z_{n-1}})$ takes the value A_{ij} if i^{th} value of the vector $\mathbf{z_n}$ is 1 and j^{th} value of the vector $\mathbf{z_{n-1}}$ is 1.
4. The generating $\mathbf{x_n}$ with $\mathbf{z_n}$ being in the particular state. This is called as emission probability and is represented as follows:

$$p(\mathbf{x_n}/\mathbf{z_n}) = \Pi_{i=1}^{i=4} (p(\mathbf{x_n}/\phi_i))^{z_{n,i}} \tag{4.45}$$

The generating probability of the sequence $p(\mathbf{x}) = p(x_1 x_2 \cdots x_N)$ using the Hidden Markov Model (Fig. 4.12a) is obtained as follows:

$$p(\mathbf{x}) = \sum_{\mathbf{z}} p(\mathbf{x}, \mathbf{z}) \tag{4.46}$$

where $p(\mathbf{x}, \mathbf{z})$ is obtained as follows:

$$\Rightarrow p(\mathbf{x}, \mathbf{z}) = p(\mathbf{z_1}) \Pi_{i=1}^{i=4} p(\mathbf{x_i}/\mathbf{z_i}) \Pi_{j=2}^{j=4} p(\mathbf{z_j}/\mathbf{z_{j-1}}) \tag{4.47}$$

$$\Rightarrow ln(p(\mathbf{x}, \mathbf{z})) = ln(p(\mathbf{z_1})) + \sum_{i=1}^{i=4} ln(\mathbf{p}(\mathbf{x_i}/\mathbf{z_i})) + \sum_{n=2}^{n=4} ln(p(\mathbf{z_n}/\mathbf{z_{n-1}})) \tag{4.48}$$

Substituting (4.43), (4.44), and (4.45) in (4.48), we get the following:

$$\sum_{i=1}^{i=4} z_{1,i} ln(\pi_i) + \sum_{i=1}^{i=4} z_{n,i} ln((p(\mathbf{x_n}/\phi_i))) + \sum_{i=1}^{i=4} \sum_{j=1}^{i=4} \sum_{n=2}^{n=4} z_{n,i} z_{n-1,j} ln(A_{ij}) \tag{4.49}$$

4.6 Expectation–Maximization (EM) Technique

The unknown hidden parameters π_k, ϕ_k, and A_{ij} are treated as follows:

$$\boldsymbol{\theta} = [\pi_k \ \phi_k \ A_{ij}] \tag{4.50}$$

Based on the observation $\mathbf{x}$, the unknown parameters are estimated by maximizing the likelihood function as follows:

$$p(\mathbf{x}/\boldsymbol{\theta}) = \sum_{\mathbf{z}} p(\mathbf{z}\mathbf{x}/\boldsymbol{\theta}) \tag{4.51}$$

Equation (4.51) is also maximized using Expectation–Maximization algorithm as follows:

1. Initialize the parameter $\boldsymbol{\theta}$ as $\boldsymbol{\theta_{old}}$.
2. The new optimal value of $\boldsymbol{\theta}$ is obtained by maximizing expectation of $ln(p(\mathbf{z}\mathbf{x}/\boldsymbol{\theta})$ over $p(\mathbf{z}/\mathbf{x}\theta_{old})$, i.e., maximizing the following:

$$\sum_{\mathbf{z}} p(\mathbf{z}/\mathbf{x}\theta_{old}) ln(p(\mathbf{z}\mathbf{x}/\boldsymbol{\theta})) \tag{4.52}$$

3. Substituting (4.49) in (4.52), we get the following three terms:

$$\sum_{\mathbf{z}} p(\mathbf{z}/\mathbf{x}\theta_{old}) \left(\sum_{i=1}^{i=4} z_{1,i} ln(\pi_i) + \sum_{i=1}^{i=4} z_{n,i} ln((p(\mathbf{x_n}/\phi_i))) \right.$$
$$\left. + \sum_{i=1}^{i=4} \sum_{j=1}^{i=4} \sum_{n=2}^{n=4} z_{n,i} z_{n-1,j} ln(A_{ij}) \right)$$

$$= \sum_{i=1}^{i=4} \sum_{\mathbf{z}} p(\mathbf{z}/\mathbf{x}\theta_{old}) z_{1,i} ln(\pi_i) + \sum_{i=1}^{i=4} \sum_{\mathbf{z}} z_{n,i} p(\mathbf{z}/\mathbf{x}\theta_{old}) ln((p(\mathbf{x_n}/\phi_i)))$$

$$+ \sum_{i=1}^{i=4} \sum_{j=1}^{i=4} \sum_{n=2}^{n=4} \sum_{\mathbf{z}} p(\mathbf{z}/\mathbf{x}\theta_{old}) z_{n,i} z_{n-1,j} ln(A_{ij})$$

$$\gamma_{z_{1,i}} = \sum_{\mathbf{z}} p(\mathbf{z}/\mathbf{x}\theta_{old}) z_{1,i} \tag{4.53}$$

$$\gamma_{z_{n,i}} = \sum_{\mathbf{z}} z_{n,i} p(\mathbf{z}/\mathbf{x}\theta_{old}) \tag{4.54}$$

$$\epsilon(n, i, n-1, j) = \sum_{\mathbf{z}} p(\mathbf{z}/\mathbf{x}\theta_{old}) z_{n,i} z_{n-1,j} \tag{4.55}$$

We get the following:

$$\sum_{\mathbf{z}} p(\mathbf{z}/\mathbf{x}\theta_{old}) ln(p(\mathbf{zx}/\boldsymbol{\theta})) \tag{4.56}$$

$$= \sum_{i=1}^{i=4} \gamma_{z_{1,i}} ln(\pi_i) + \sum_{i=1}^{i=4} \gamma_{z_{n,i}} ln((p(\mathbf{x_n}/\phi_i)))$$

$$+ \sum_{i=1}^{i=4} \sum_{j=1}^{i=4} \sum_{n=2}^{n=4} \epsilon(n, i, n-1, j) ln(A_{ij}) \tag{4.57}$$

It is observed that given $\mathbf{x}$, the probability that the first element being generated from the ith state is given as $\gamma_{z_{1,i}}$. Similarly given $\mathbf{x}$, the probability that the nth element being generated from the ith state is given as $\gamma_{z_{n,i}}$. They are called as responsibility factor. In the same fashion, given $\mathbf{x}$, the probability that n^{th} element being generated from the ith state, provided $(n-1)^{th}$ element being generated from the jth state is given as $\epsilon(n, i, n-1, j)$.

4.6.1 *Expectation Stage*

The new set of values (expected values) for $\gamma_{z_{n,i}}$, $\epsilon(n, i, n-1, j)$, and $\gamma_{z_{1,i}}$ is obtained in this stage using $\alpha(\mathbf{z_n})$ and $\beta(\mathbf{z_n})$ as follows:

$$\gamma_{\mathbf{z_n}} = p(\mathbf{z_n}/\mathbf{x}) \tag{4.58}$$

$$= \frac{p(\mathbf{x}/\mathbf{z_n}) p(\mathbf{z_n})}{p(\mathbf{x})} \tag{4.59}$$

$$= \frac{p(\mathbf{x_1}\ \mathbf{x_2}\ \mathbf{x_n}/\mathbf{z_n})p(\mathbf{x_{n+1}}\ \mathbf{x_{n+2}}\ \cdots \mathbf{x_N}/\mathbf{z_n})p(\mathbf{z_n})}{p(\mathbf{x})} \tag{4.60}$$

$$\frac{p(\mathbf{x_1}\ \mathbf{x_2}\ \mathbf{x_n}, \mathbf{z_n})p(\mathbf{x_{n+1}}\ \mathbf{x_{n+2}}\ \cdots \mathbf{x_N}/\mathbf{z_n})}{p(\mathbf{x})} \tag{4.61}$$

$$\gamma_{\mathbf{z_n}} = \frac{\alpha(\mathbf{z_n})\beta(\mathbf{z_n})}{p(\mathbf{x})} \tag{4.62}$$

Similarly, $\epsilon(n, n-1) = p(\mathbf{z_n}, \mathbf{z_{n-1}}/\mathbf{x})$ is computed as follows:

$$\begin{aligned}
\epsilon(n, n-1) &= p(\mathbf{z_n}, \mathbf{z_{n-1}}/\mathbf{x}) \\
&= \frac{p(\mathbf{x}/\mathbf{z_n}, \mathbf{z_{n-1}})p(\mathbf{z_n}, \mathbf{z_{n-1}})}{p(\mathbf{x})} \\
&= \frac{p(\mathbf{x_1}\mathbf{x_2}\cdots \mathbf{x_n}\mathbf{x_{n+1}}\cdots \mathbf{x_N}/\mathbf{z_n}, \mathbf{z_{n-1}})p(\mathbf{z_n}/\mathbf{z_{n-1}})p(\mathbf{z_{n-1}})}{p(\mathbf{x})} \\
&= \frac{p(\mathbf{x_1}\mathbf{x_2}\cdots \mathbf{x_{n-1}}/\mathbf{z_n}, \mathbf{z_{n-1}})p(\mathbf{x_n}/\mathbf{z_n}, \mathbf{z_{n-1}})p(\mathbf{x_{n+1}}\cdots \mathbf{x_N}/\mathbf{z_n}, \mathbf{z_{n-1}})p(\mathbf{z_n}/\mathbf{z_{n-1}})p(\mathbf{z_{n-1}})}{p(\mathbf{x})} \\
&= \frac{\alpha(\mathbf{z_{n-1}})p(\mathbf{x_n}/\mathbf{z_n})\beta(\mathbf{z_n})p(\mathbf{z_n}/\mathbf{z_{n-1}})}{p(\mathbf{x})}
\end{aligned}$$

$\alpha(\mathbf{z_n})$ is recursively computed as follows:

$$\alpha(\mathbf{z_n}) = p(\mathbf{x_1}\ \mathbf{x_2}\ \cdots \mathbf{x_n}, \mathbf{z_n}) \tag{4.63}$$

$$= p(\mathbf{x_1}\ \mathbf{x_2}\ \cdots \mathbf{x_{n-1}}, \mathbf{x_n}/\mathbf{z_n})p(\mathbf{z_n}) \tag{4.64}$$

$$= p(\mathbf{x_1}\ \mathbf{x_2}\ \cdots \mathbf{x_{n-1}}/\mathbf{z_n})p(\mathbf{z_n})p(\mathbf{x_n}/\mathbf{z_n}) \tag{4.65}$$

$$= p(\mathbf{x_1}\ \mathbf{x_2}\ \cdots \mathbf{x_{n-1}}, \mathbf{z_n})p(\mathbf{x_n}/\mathbf{z_n}) \tag{4.66}$$

$$= p(\mathbf{x_n}/\mathbf{z_n})\sum_{z_{n-1}} p(\mathbf{x_1}\ \mathbf{x_2}\ \mathbf{x_{n-1}}, \mathbf{z_{n-1}}, \mathbf{z_n}) \tag{4.67}$$

$$= p(\mathbf{x_n}/\mathbf{z_n})\sum_{z_{n-1}} p(\mathbf{x_1}\ \mathbf{x_2}\ \mathbf{x_{n-1}}, \mathbf{z_n}/\mathbf{z_{n-1}})p(\mathbf{z_{n-1}}) \tag{4.68}$$

$$= p(\mathbf{x_n}/\mathbf{z_n})\sum_{z_{n-1}} p(\mathbf{x_1}\ \mathbf{x_2}\ \mathbf{x_{n-1}}, \mathbf{z_n}/\mathbf{z_{n-1}})p(\mathbf{z_{n-1}}) \tag{4.69}$$

$$= p(\mathbf{x_n}/\mathbf{z_n})\sum_{z_{n-1}} p(\mathbf{x_1}\ \mathbf{x_2}\ \mathbf{x_{n-1}}, /\mathbf{z_{n-1}})p(\mathbf{z_{n-1}})p(\mathbf{z_n}/\mathbf{z_{n-1}}) \tag{4.70}$$

$$= p(\mathbf{x_n}/\mathbf{z_n})\sum_{z_{n-1}} p(\mathbf{x_1}\ \mathbf{x_2}\ \mathbf{x_{n-1}}, \mathbf{z_{n-1}})p(\mathbf{z_n}/\mathbf{z_{n-1}}) \tag{4.71}$$

$$= p(\mathbf{x_n}/\mathbf{z_n})\sum_{z_{n-1}} \alpha(\mathbf{z_{n-1}})p(\mathbf{z_n}/\mathbf{z_{n-1}}) \tag{4.72}$$

$\beta(\mathbf{z_n})$ is recursively computed as follows:

$$\beta(\mathbf{z_n}) = p(\mathbf{x_{n+1}}\ \mathbf{x_{n+2}}\ \cdots \mathbf{x_N}/\mathbf{z_n}) \tag{4.73}$$

$$= \sum_{\mathbf{z_{n+1}}} p(\mathbf{x_{n+1}}\ \mathbf{x_{n+2}}\ \cdots \mathbf{x_N}, \mathbf{z_{n+1}}/\mathbf{z_n}) \tag{4.74}$$

$$= \sum_{\mathbf{z_{n+1}}} \frac{p(\mathbf{x_{n+1}}\ \mathbf{x_{n+2}}\ \cdots \mathbf{z_n}\mathbf{z_{n+1}})}{p(\mathbf{z_n})} \tag{4.75}$$

$$= \sum_{\mathbf{z_{n+1}}} \frac{p(\mathbf{x_{n+1}}\ \mathbf{x_{n+2}}\ \cdots \mathbf{z_n}/\mathbf{z_{n+1}})p(\mathbf{z_{n+1}})}{p(\mathbf{z_n})} \tag{4.76}$$

$$= \sum_{\mathbf{z_{n+1}}} \frac{p(\mathbf{x_{n+1}}/\mathbf{z_{n+1}})p(\mathbf{x_{n+2}}\cdots \mathbf{x_N}/\mathbf{z_{n+1}})p(\mathbf{z_n}/\mathbf{z_{n+1}})p(\mathbf{z_{n+1}})}{p(\mathbf{z_n})} \tag{4.77}$$

$$= \sum_{\mathbf{z_{n+1}}} \frac{p(\mathbf{x_{n+1}}/\mathbf{z_{n+1}})\beta(\mathbf{z_{n+1}})p(\mathbf{z_n}/\mathbf{z_{n+1}})p(\mathbf{z_{n+1}})}{p(\mathbf{z_n})} \tag{4.78}$$

$$= \sum_{\mathbf{z_{n+1}}} p(\mathbf{x_{n+1}}/\mathbf{z_{n+1}})\beta(\mathbf{z_{n+1}})p(\mathbf{z_{n+1}}/\mathbf{z_n}) \tag{4.79}$$

4.6.2 Maximization Stage

The unknown parameters $\boldsymbol{\theta}$ are obtained by maximizing (4.56) in terms of $\gamma_{z_{n,i}}$, $\epsilon(n, i, n-1, j)$, and $\gamma_{z_{1,i}}$.

1. Optimizing π_l with the constraint $\sum_{i=1}^{i=K} \pi_i = 1$ (with K states) is obtained as follows. The Lagrange equation is as follows:

$$\sum_{i=1}^{i=K} \gamma_{z_{1,i}} ln(\pi_i) + \sum_{i=1}^{i=K} \gamma_{z_{n,i}} ln((p(\mathbf{x_n}/\phi_i))) + \sum_{i=1}^{i=K}\sum_{j=1}^{i=K}\sum_{n=2}^{n=K} \epsilon(n, i, n-1, j)$$
$$\times ln(A_{ij}) - \lambda\left(\sum_{k=1}^{k=K} \pi_k - 1\right)$$

Differentiating the above equation with respect to π_l and equating to zero, we get the following:

$$\frac{\gamma_{z_{1,l}}}{(\pi_l)} - \lambda = 0$$
$$\Rightarrow \lambda = \frac{\gamma_{z_{1,l}}}{(\pi_l)}$$

$$\Rightarrow \sum_{r=1}^{r=K} \pi_r \lambda = \sum_{r=1}^{r=K} \gamma_{z_{1,r}}$$

$$\Rightarrow \lambda = \sum_{r=1}^{r=K} \gamma_{z_{1,r}}$$

$$\Rightarrow \pi_l = \frac{\gamma_{z_{1,l}}}{\sum_{r=1}^{r=K} \gamma_{z_{1,r}}}$$

2. $A_{i,j} = \frac{\sum_{n=2}^{n=N} \epsilon(n,i,n-1,j)}{\sum_{i=1}^{i=K} \sum_{n=2}^{n=N} \epsilon(n,i,n-1,j)}$
3. If $p(\mathbf{x_n}/\phi_i)$ is assumed as multivariate Gaussian density function, the mean vector and the co-variance matrix are computed as follows:

$$\mu_k = \frac{\sum_{n=1}^{n=N} \gamma_{z_{n,k}} X_n}{\sum_{n=1}^{n=N} \gamma_{z_{n,k}}}$$

$$\sum_k = \frac{\sum_{n=1}^{n=N} \gamma_{z_{n,k}} (\mathbf{x_n} - \mu_k)((\mathbf{x_n} - \mu_k))^T}{\sum_{n=1}^{n=N} \gamma_{z_{n,k}}}$$

hmmgenerateseq.m

```
%Data generation
function [state,DATA]=hmmgenerateseq(p11,p21,h11,h21,
prior)
DATA=[];
if(rand<prior)
state=1;
else
state=2;
end
for i=1:1:50000
if(state(i)==1)
    if(rand<p11)
        state=[state 1];
    else
        state=[state 2];
    end
else
    if(rand<p21)
        state=[state 1];
    else
        state=[state 2];
    end
end
```

```
if(state(i)==1)
    if(rand<h11)
        DATA=[DATA 1];
    else
        DATA=[DATA 2];
    end
        else
    if(rand<h21)
        DATA=[DATA 1];
    else
        DATA=[DATA 2];
    end
        end
end
end
```

genprob.m

```
function [prob]=genprob(seq,p11,p21,h11,h21)
s=1;
prob=0;
for i=1:1:length(seq)
    if(seq(i)==1)
        if(s==1)
    choice1=prob+log(p11)+log(h11);
    choice2=prob+log((1-p11))+log(h21);
        else
    choice1=prob+log(p21)+log(h11);
    choice2=prob+log((1-p21))+log(h21);
        end
    else
    if(s==1)
    choice1=prob+log(p11)+log((1-h11));
    choice2=prob+log((1-p11))+log((1-h21));
        else
    choice1=prob+log(p21)+log((1-h11));
    choice2=prob+log((1-p21))+log((1-h21));
    end
    end
if(choice1>choice2)
    prob=choice1;
s=1;
else
    prob=choice2;
s=2;
end
```

```
end
end
```

sequencehmm.m

```
%Check for the sequence pattern based on the
constructed HMM model.
%sequencehmm.m
p11=0.8; p21=0.3;
h11=0.9;h21=0.2;
[state1,DATA1]=hmmgenerateseq(p11,p21,h11,h21,prior);
q11=0.2; q21=0.7;
r11=0.1; r21=0.8;
[state2,DATA2]=hmmgenerateseq(q11,q21,r11,r21,prior);
g1=genprob(DATA1,p11,p21,h11,h21);
g2=genprob(DATA1,q11,q21,r11,r21);
g3=genprob(DATA2,p11,p21,h11,h21);
g4=genprob(DATA2,q11,q21,r11,r21);

if(g1>g2)
    display('DATAT 1 is from model 1')
else
    display('DATAT 1 is from model 2')
end

if(g3>g4)
    display('DATAT 2 is from model 1')
else
    display('DATAT 2 is from model 2')
end
```

generatehmmmodel.m

```
%generatehmmmodel.m
%p's are state transition probabilities
%and h are generation probabilities
%p11->transition from 1 to 1
%p12->transition from 1 to 2
%p21->transition from 2 to 1
%p22->transition from 2 to 2
%h11->being in the state 1, 1 is generated
%h12->being in the state 1, 2 is generated
%h21->being in the state 2, 1 is generated
%h22->being in the state 2, 2 is generated
prior=1;
p11=0.4;
p12=1-p11;
p21=0.6;
```

```
p22=1-p21;
h11=0.3;
h12=1-h11;
h21=0.5;
h22=1-h21;
REF=[prior p11 p21 h11 h21];
[state,data]=hmmgenerateseq(p11,p21,h11,h21,prior);
%Given the sequence data with state as hidden,
%HMM is obtained as follows.
%Initialize prior,p11,p21,h11,h21 and
%the intial state is set to 1
%h11 and h21 are the generating probabilities
%p11 and p21 are the state transition probabilities
prior=1;
p11=rand;
p12=1-p11;
p21=rand;
p22=1-p21;
h11=rand;
h12=1-h11;
h21=rand;
h22=1-h21;
%Expectation stage
%In this r(n,i) e(n,i,n-1,j) and r(1,i) are obtained.
%Given the sequence, r(n,i) gives the responsibility
factor that
%nth frame being generated with ith state
%e(n,i,n-1,j) gives the probability that nth frame
generated
%from the ith state and (n-1)th frame generated from
the jth state.
%Maximization stage gives the actual state transition
probability
%and the generating probability
%alpha(n,k) is the probability p(x1 x2 ... xn,znk)
%(i.e) nth frame generated from kth state
PX=[];
COL=[];
for iter=1:1:50
if(data(1)==1)
alpha(1,1)=prior*h11;
alpha(1,2)=(1-prior)*h21;
else
    alpha(1,1)=prior*h12;
    alpha(1,2)=(1-prior)*h22;
```

```
end

for n=2:1:length(data)
    if(data(n)==1)
    alpha(n,1)=h11*(alpha(n-1,1)*p11+alpha(n-1,2)
    *p21);
    alpha(n,2)=h21*(alpha(n-1,1)*p12+alpha(n-1,2)
    *p22);
    else
    alpha(n,1)=h12*(alpha(n-1,1)*p11+alpha(n-1,2)
    *p21);
    alpha(n,2)=h22*(alpha(n-1,1)*p12+alpha(n-1,2)
    *p22);
    end
end

%beta(n,k) is the probability p(xn+1 xn+2 ... xN/Znk)
%beta(N,1) and beta(N,2) are considered as 1
beta(length(data),1)=1;
beta(length(data),2)=1;
for m=length(data)-1:-1:1
    if(data(m+1)==1)
    beta(m,1)=(h11*beta(m+1,1)*p11+h21*beta(m+1,2)
    *p12);
    beta(m,2)=(h11*beta(m+1,1)*p21+h21*beta(m+1,2)
    *p22);
    else
    beta(m,1)=(h12*beta(m+1,1)*p11+h22*beta(m+1,2)
    *p12);
    beta(m,2)=(h12*beta(m+1,1)*p21+h22*beta(m+1,2)
    *p22);
    end
end
%probability of data generation is represented as px
%computation of gamma
px=alpha(length(data),1)*beta(length(data),1)+...
alpha(length(data),2)*beta(length(data),2);
for n=1:1:length(data)
gamma(n,1)=(alpha(n,1)*beta(n,1))/(px);
gamma(n,2)=(alpha(n,2)*beta(n,2))/(px);
end
%computation of e(n,i,j) (nth data generated from
ith state
%and (n-1) the data genererted from jth state.
for n=2:1:length(data)
```

```
    if(data(n)==1)
    e(n,1,1)=alpha(n-1,1)*h11*beta(n,1)*p11;
    e(n,1,2)=alpha(n-1,2)*h11*beta(n,1)*p21;
    e(n,2,1)=alpha(n-1,1)*h21*beta(n,2)*p12;
    e(n,2,2)=alpha(n-1,2)*h21*beta(n,2)*p22;
    else
    e(n,1,1)=alpha(n-1,1)*h12*beta(n,1)*p11;
    e(n,1,2)=alpha(n-1,2)*h12*beta(n,1)*p21;
    e(n,2,1)=alpha(n-1,1)*h22*beta(n,2)*p12;
    e(n,2,2)=alpha(n-1,2)*h22*beta(n,2)*p22;
    end
end
%Maximization stage (Estimate p11,p12,h11,h12,
prior...
%using e,gamma
prior=gamma(1,1)/(gamma(1,1)+gamma(1,2));
num1=0;
den1=0;
num2=0;
den2=0;
for n=2:1:length(data)
num1=num1+e(n,1,1);
num2=num2+e(n,1,2);
for i=1:2
den1=den1+e(n,i,1);
den2=den2+e(n,i,2);
end
end
p11=(num1)/(den1);
p21=(num2)/(den2);
[u v]=find(data==1);
h11=sum(gamma(v,1))/sum(gamma(:,1));
h12=1-h11;
h21=sum(gamma(v,2))/sum(gamma(:,2));
h22=1-h21;
COL{iter}=[prior p11 p21 h11 h21];
PX=[PX px];
end
[p,q]=max(gamma');
length(find((data-q)==0))
display("'ORIGINAL probabilities of the HMM Model')
REF
display('Probabilities obtained using EM algorithm')
COL{iter}
figure
```

```
plot(PX)
xlabel('Iterations')
ylabel('Generating probability')
```

4.6.3 *Illustration of Hidden Markov Model*

In this illustration, the sequence of binary sequence (represented as 1 (green) and 2 (orange)) generated using Hidden Markov Model is considered (refer Fig. 4.13). The number of states is considered as 2. Being in the particular state (say i), the data 1 or data 2 is generated with the generating probability given as p_{i1} and p_{i2}, respectively. Also the state transition probability h_{ij} describes the change in the state from i to j (Figs. 4.14 and 4.15). The sequence of actions to generate the binary sequence is as follows:

1. Initialize the state (says s) based on the prior probability of the state $prior$.

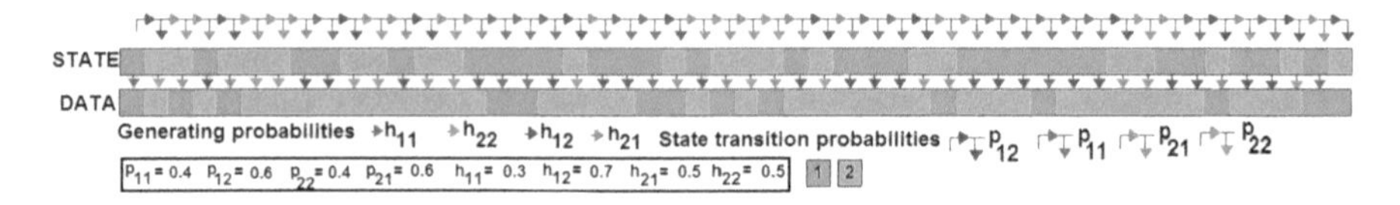

Fig. 4.13 Illustration of binary sequence generation using Hidden Markov Model. The model is described by the state transition probabilities, generating probabilities, and the prior probability

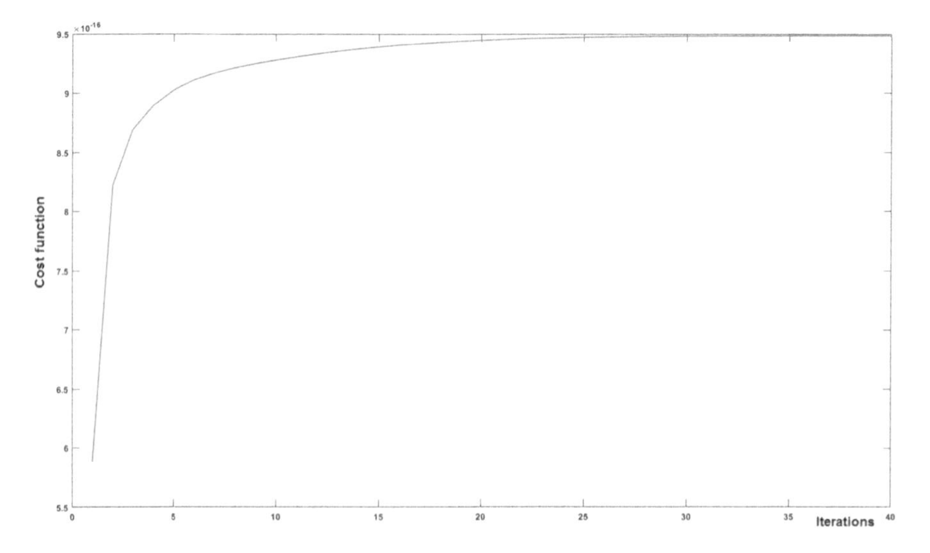

Fig. 4.14 Illustration of maximization of the generating probability by optimizing HMM using Expectation–Maximization algorithm

Probability details	Prior	p_{11}	p_{21}	h_{11}	h_{21}
Iteration 1	1.0000	0.0828	0.3839	0.5225	0.3958
Iteration50	1.0000	0.2679	0.5457	0.4631	0.4568
Actual	1.0000	0.4000	0.6000	0.3000	0.5000

Fig. 4.15 Illustration of convergence obtained using Expectation–Maximization algorithm to HMM

2. Being in the state s, generate the single binary data 1 or 2 based on the generating probability p_{s1} and p_{s2}.
3. After generating the single binary data, the state transition from s to t occurs based on the state transition probability h_{s1} and h_{s2}.
4. Update s with t.
5. Repeat (2)–(4) to generate the binary sequence.

In the real time applications, the model (estimation of $prior, p_{11}, p_{12}, p_{21}, p_{22}, h_{11}, h_{12}, h_{21}, h_{22}$) needs to be obtained based on the observation of the binary sequence. These are obtained as follows:

1. The probability that the binary sequence $[x_1\ x_2 \cdots x_n]$ with n^{th} binary digit being generated from the kth state is represented as $\alpha(n,k)$. Similarly, probability that the binary sequence $[x_{n+1}\ x_{n+2} \cdots x_N]$ conditioned on n^{th} binary digit being generated from the kth state is represented as $\beta(n,k)$ (refer (4.62) and (4.72)).
2. Initialize the model parameters.
3. $\alpha(n,k)$ for $n = 1 \cdots N$ and $k = 1 \cdots N$ are computed.
4. Check for the first data. If it is 1, $\alpha(1,1) = prior * h_{11}$ and $\alpha(1,2) = (1 - prior) * h_{21}$. $*$ is the multiplication operator.
5. If the first data 2, $\alpha(2,1) = prior * h_{12}$ and $\alpha(2,2) = (1 - prior) * h_{22}$.
6. Check for the next data (n). If it is 1, $\alpha(n,1) = \alpha(n-1,1) * h11 * p11 + \alpha(n-1,2) * h11 * p21$, and $\alpha(n,2) = \alpha(n-1,1) * h21 * p12 + \alpha(n-1,2) * h21 * p22$.
7. Check for the next data (n). If it is 2, $\alpha(n,1) = \alpha(n-1,1) * h12 * p11 + \alpha(n-1,2) * h12 * p21$, and $\alpha(n,2) = \alpha(n-1,1) * h22 * p12 + \alpha(n-1,2) * h22 * p22$.
8. Repeat steps (6)–(7) for $n = 1 \cdot N$ and $k = 1 \cdots N$.
9. Initialize $\beta(N,1) = 1$ and $\beta(N,2) = 1$.
10. Check for the last data (N). If it is 1, $\beta(N-1,1) = h_{11} * \beta(N,1) * p_{11} + h_{21} * \beta(N,1) * p_{12}$ and $\beta(N-1,2) = h_{11} * \beta(N,1) * p_{21} + h_{21} * \beta(N,1) * p_{22}$.
11. If it is 2, $\beta(N-1,1) = h_{12} * \beta(N,1) * p_{11} + h_{22} * \beta(N,1) * p_{12}$ and $\beta(N-1,2) = h_{12} * \beta(N,1) * p_{21} + h_{22} * \beta(N,1) * p_{22}$.
12. Check for the next data from the last (m). $\beta(m-1,1) = h_{11} * \beta(m,1) * p_{11} + h_{21} * \beta(m,1) * p_{12}$ and $\beta(m-1,2) = h_{11} * \beta(m,1) * p_{21} + h_{21} * \beta(m,1) * p_{22}$.
13. If it is 2, $\beta(m-1,1) = h_{12} * \beta(m,1) * p_{11} + h_{22} * \beta(m,1) * p_{12}$ and $\beta(m-1,2) = h_{12} * \beta(m,1) * p_{21} + h_{22} * \beta(m,1) * p_{22}$.
14. Repeat the steps (12) and (13) for $n = 1 \cdots N$ and $k = 1 \cdots N$ are computed.
15. Compute the generating probability of the complete data as $px = \alpha(N,1)\beta(N,1) + \alpha(N,2)\beta(N,2)$.
16. Compute the responsibility factor $\gamma(n,k)$, n^{th} data generated from the kth state as follows: $\gamma(n,k) = \frac{\alpha(n,k)\beta(n,k)}{px}$

17. Compute the responsibility factor $e(n, p, q)$ (n^{th} data generated from the pth state and $(n-1)^{th}$ data generated from q^{th} state).
18. Check for the second data. If it is 1, $e(n, 1, 1) = \alpha(n-1, 1) * h_{11} * \beta(n, 1) * p_{11}$,$e(n, 1, 2) = \alpha(n-1, 2) * h_{11} * \beta(n, 1) * p_{21}$,$e(n, 2, 1) = \alpha(n-1, 1) * h_{21} * \beta(n, 2) * p_{12}$,$e(n, 2, 2) = \alpha(n-1, 2) * h_{21} * \beta(n, 2) * p_{22}$.
19. If it is 2, $e(n, 1, 1) = \alpha(n-1, 1) * h_{12} * \beta(n, 1) * p_{11}$,$e(n, 1, 2) = \alpha(n-1, 2) * h_{12} * \beta(n, 1) * p_{21}$,$e(n, 2, 1) = \alpha(n-1, 1) * h_{22} * \beta(n, 2) * p_{12}$,$e(n, 2, 2) = \alpha(n-1, 2) * h_{22} * \beta(n, 2) * p_{22}$.
20. Repeat steps (3) and (19) for all the data up to N.
21. This completes the expectation stage. Using the estimated values obtained during the expectation stage, the unknown hidden parameters are estimated by maximizing the generating probability. This is done as the maximization stage.
22. $prior = \frac{\gamma(1,1)}{\gamma(1,1)+\gamma(1,2)}$
23. $p_{11} = \frac{\sum_{n=2}^{N} e(n,1,1)}{\sum_{n=2}{n=N} \sum_{i=1}^{i=2} e_{n,i,1}}$
24. $p_{21} = \frac{\sum_{n=2}^{N} e(n,1,2)}{\sum_{n=2}{n=N} \sum_{i=1}^{i=2} e_{n,i,2}}$
25. Compute the summation of the $\gamma(m, 1)$ for all m corresponding to the $data\ value = 1$. Let it be num. Compute the summation of the $\gamma(m, 1)$ for all m. Let it be den. Declare it as $h_{11} = \frac{num}{den}$ and $h_{12} = 1 - h_{11}$.
26. Compute the summation of the $\gamma(m, 2)$ for all m corresponding to the $data\ value = 1$. Let it be num. Compute the summation of the $\gamma(m, 2)$ for all m. Let it be den. Declare it as $h_{21} = \frac{num}{den}$. $h_{22} = 1 - h_{21}$
27. Go to step (3). Repeat (3)–(26) for finite number of iterations or until probability of generating (px) converges to the larger value.
28. Thus HMM is obtained.

4.7 Unsupervised Clustering

4.7.1 k-Means Algorithm

Given the data, we would like to classify them into finite number of groups. This is known as unsupervised clustering (Fig. 4.16). This can be done using k-means as follows:

1. Initialize the number of classes (K) and the corresponding mean vectors ($\boldsymbol{\mu_k}$ for $k = 1 \cdots K$). Let it be K.
2. Assign the element $t_{nk} = 1$ corresponding to the vector $\mathbf{x_n}$ if

$$k = arg_{i=1}^{i=K} min \sum_{n=1}^{n=N} (\mathbf{x_n} - \mu_i)^2. \tag{4.80}$$

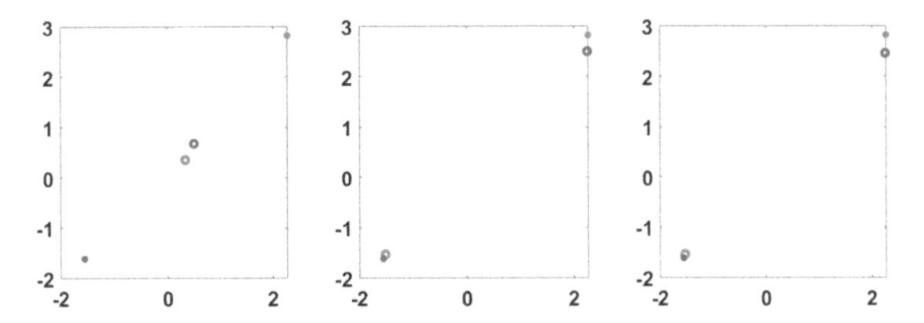

Fig. 4.16 Illustration of convergence of mean using k-means algorithm

3. The mean vectors are obtained by minimizing J as shown in the following:

$$J = \sum_{k=1}^{k=K} \sum_{n=1}^{n=N} (x_n t_{nk} - \mu_k t_{nk})^2 \tag{4.81}$$

where t_{nk} is 1 if n^{th} vector is assigned to k^{th} class. Differentiating (4.81) with respect to μ_{nk} and equating to zero, we get the following:

$$\mu_k = \frac{\sum_{k=1}^{k=K} \sum_{n=1}^{n=N} x_n t_{nk}}{\sum_{k=1}^{k=K} \sum_{n=1}^{n=N} t_{nk}} \tag{4.82}$$

4. Repeat (2) and (3) until convergence occurs.

kmeans.m

```
%kmeans algorithm
%Generation of data
C1=[0.7 0.3;0.2 0.9]
[E1,D1]=eig(C1);
M1=rand(1,2)+[2 2];
C2=[0.9 0.1;0.4 0.6];
[E2,D2]=eig(C2);
M2=rand(1,2)+[-2 -2];
%The prior probability of the first Mixture and the
second Mixtures are
%0.2 and 0.8 respectively.
%Generation of 10000 data with additive noise from
constructed Model is as
%given below
figure(1)
DATA=[]
INDEX=[];
for l=1:1:200
```

```
r=rand;
if(r<0.2)
    temp=randn(2,1);
    temp1=E1'*D1.^(1/2)*temp+M1;
    hold on
    plot(temp1(1),temp1(2),'ro');
    %second model
DATA=[DATA temp1];
INDEX=[INDEX 1];
else
    %first model
temp=randn(2,1);
temp2=E2'*D2.^(1/2)*temp+M2;
hold on
plot(temp2(1),temp2(2),'bo');
DATA=[DATA temp2];
INDEX=[INDEX 2];
end
end
%DATA is clustered using kmeans algorithm as
given below.
%Initialize mean
m1=rand(2,1);
m2=rand(2,1);
M{1}=[m1 m2];
for iter=2:1:100
for i=1:1:length(DATA)
[P1]=sum((DATA-repmat(m1,1,400)).^2) ;
[P2]=sum((DATA-repmat(m2,1,400)).^2) ;
end
P=[P1;P2];
[U,V]=min(P);
[x1,y1]=find(V==1);
[x2,y2]=find(V==2);
m1=mean(DATA(:,y1)')';
m2=mean(DATA(:,y2)')';
temp1=DATA(:,V(y1));
temp2=DATA(:,V(y2));
TEMP1{iter}=temp1;
TEMP2{iter}=temp2;
M{iter}=[m1 m2];
end
t=[1 2 3];
for i=1:1:3
subplot(1,3,i)
```

```
plot(M{t(i)}(1,1),M{t(i)}(2,1),'ro')
hold on
plot(M1(1,1),M1(1,2),'r*')
hold on
plot(M{t(i)}(1,2),M{t(i)}(2,2),'bo')
hold on
plot(M2(1,1),M2(1,2),'b*')
end
```

4.7.2 GMM-Revisited

1. Suppose if we have the data belonging to two different classes. We can obtain the GMM (with number of mixtures arbitrarily chosen) for the individual classes (say GMM1 and GMM2) based on the training set. Once it is done, to classify the unknown data into one among the two classes, we obtain the generating probability computed using models 1 and 2. We declare the vector belongs to class 1 if the generating probability computed using model 1 is larger.
2. Given set of vectors, unsupervised clustering (i.e., grouping the vectors into K number of groups) is performed using GMM. Once GMM is obtained, we classify the vector into Group r if

$$r = arg_k max_{k=1}^{k=K} p(c_i)p(x/c_i) \tag{4.83}$$

It is noted in this case that the class index for the individual training sets was not known.

4.8 Summary

1. The probabilistic approach of classification identifies the class conditional pdf $p(c_k/\mathbf{x})$ for $k = 1 \cdots r$ and identifies the index for which the ccpdf is maximum (Fig. 4.17).
2. The class conditional pdf is modeled as sigmoidal function. In case of multi-class problem, soft max function is used to represent ccpdf. The gradient and the Newton–Raphson methods are used to optimize the unknown vector **w** used in the function.
3. For the two-class problem, the sigmoidal function $f(w^T\phi(\mathbf{x})) = \frac{1}{1+e^{-w^T\phi(\mathbf{x})}}$ is used. Instead probit function (cumulative distribution of Gaussian density function with mean 0 and variance 1) can also be as the $f(w^T\phi(\mathbf{x}))$.
4. In the case of Bayesian logistic regression, the **w** used in the logistic regression is assumed as the Gaussian distributed with mean $\mathbf{m_0}$ and co-variance matrix $\mathbf{S_0}$.

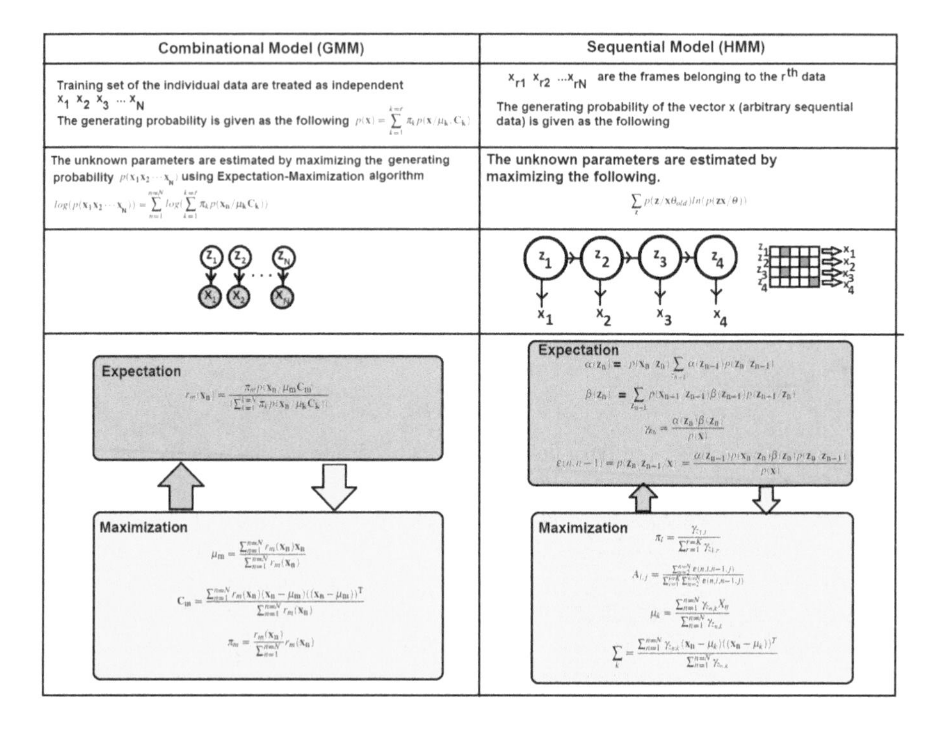

Fig. 4.17 Hidden Markov Model versus Gaussian Mixture Model

5. The posterior density function of over **w** is given as

$$-\frac{1}{2}(w-m_0)^T S_0^{-1}(w-m_0)+\sum_{n=1}^{n=N} t_n ln(t_n)+(1-t_n)ln(1-y_n)+constant \tag{4.84}$$

6. The posterior density function is approximated as the Gaussian distributed with mean vector at $\nabla p(\mathbf{w}/t)=0$, which is given as follows:

$$\Phi^T(t-y)-S_0^{-1}\mathbf{w}=0\mathbf{w}=S_0\Phi^T(t-y) \tag{4.85}$$

7. The co-variance matrix is approximated as $(\Phi^T B\Phi+S_0)^{-1}$, where the diagonal elements of the matrix B is filled up with $y_n(1-y_n)$.
8. The optimal **w** is obtained by iteratively as follows:
   (a) Initialize the diagonal elements of S_0 as $\frac{1}{\alpha_i}$
   (b) Compute the co-variance matrix $(\Phi^T B\Phi+S_0)^{-1}$
   (c) Compute $\alpha_i=\frac{\gamma_i}{w_i^2}$
   (d) Repeat the steps to obtain the optimal vector **w**

9. Also for the two-class logistic regression, the approximate predictive probability is obtained using the sigmoid function as follows:

$$p(c_1/\boldsymbol{\phi}(\mathbf{x})) = \frac{1}{1 + e^{-(1+\pi\sigma_a^2/8)^{-1/2}\mu_a}} \tag{4.86}$$

where σ_a and μ_a are the variance and the mean of the $\mathbf{w}^T\phi(\mathbf{x})$ with prior density function of $\mathbf{w}$ as Gaussian distributed with mean zero and co-variance matrix S_0.

Chapter 5
Computational Intelligence

5.1 Particle Swarm Optimization (PSO) to Maximize Gaussianity

The algorithm inspired from the birds' (particle) strategy in reaching the destination is adopted to formulate the Particle Swarm Optimization (refer Fig. 5.1). Let the position of the particle be represented as $\mathbf{w}$. The distance from the destination is the absolute of the kurtosis (zero in case of Gaussian data). The algorithm is formulated to obtain the transformation matrix such that the absolute value of the kurtosis is minimized as follows:

1. Initialize the positions of the birds. Let it be $\mathbf{x_1}, \mathbf{x_2}, \mathbf{x_3}, \mathbf{x_4}, \ldots, \mathbf{x_N}$. Let the corresponding functional values are computed as $\mathbf{f(x_1)}, \mathbf{f(x_2)}, \mathbf{f(x_3)}, \ldots, \mathbf{f(x_N)}$.
2. Let the individual birds are trying to move towards the destination. These are treated as the tentative local decisions. Let it be $\mathbf{y_1}, \mathbf{y_2}, \mathbf{y_3}, \mathbf{y_4}, \ldots, \mathbf{y_N}$. Let the corresponding functional values are computed as follows: $\mathbf{f(y_1)}, \mathbf{f(y_2)}, \mathbf{f(y_3)}, \ldots,$ $\mathbf{f(y_N)}$.
3. The one among the N particle (local decisions) is declared as the Global decision. This index of the Global decision is obtained as $g = arg_k min f(\mathbf{y_k})$.
4. The actual movement of the individual particles is obtained as follows: $z_k = x_k + \alpha \times (y_k - x_k) + \beta \times (y_g - x_k)$. This completes one iteration.
5. Now z_k becomes the current positions, i.e., $x_k = z_k$. The tentative next move of the kth particle is obtained as the one that gives the minimum among $f(y_k)$ and $g(z_k)$.
6. Repeat (3)–(5) for finite number of iterations (until convergence occurs).
7. Track the best particle (that gives the minimum functional value) in each iteration and the best among the track is declared as the solution. The illustration on maximizing the Gaussianity using PSO is illustrated in Figs. 5.2 and 5.3.

© Springer Nature Switzerland AG 2020

E. S. Gopi, *Pattern Recognition and Computational Intelligence Techniques Using Matlab*, Transactions on Computational Science and Computational Intelligence,
https://doi.org/10.1007/978-3-030-22273-4_5

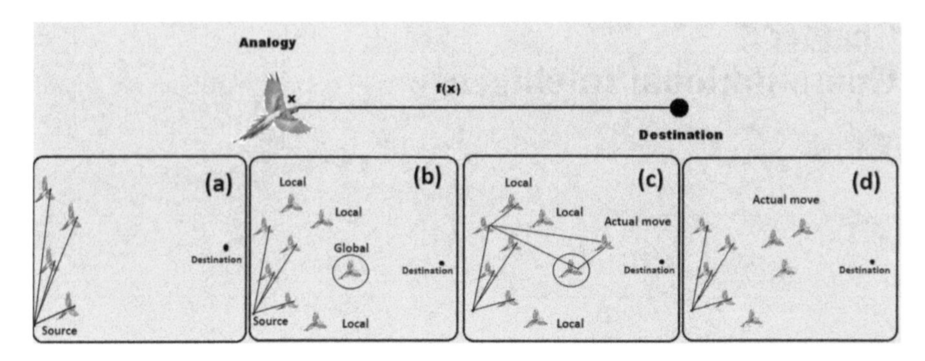

Fig. 5.1 (**a**) Initial positions of the swarms. (**b**) Tentative next positions selected by the swarms (with global marked with the circle). (**c**) Actual movement as the linear combinations of local decision and global decision. (**d**) Initial positions and the actual move

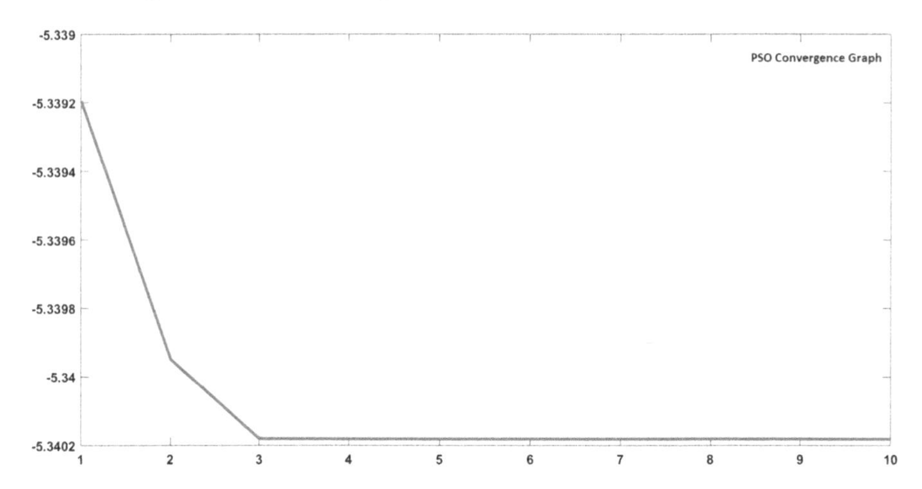

Fig. 5.2 Convergence of PSO to maximize Gaussianity by minimizing the absolute value of the kurtosis

psokurtosis.m

```
%Particle Swarm Optimization to minimize
%the absolute value of the kurtosis
%Initializing 100 particles
load x
x=x-repmat(mean(x')',1,length(x));
figure
subplot(1,2,1)
plot(x(1,:),x(2,:),'r*')
VAL1=[];
VAL2=[];
for i=1:1:2
```

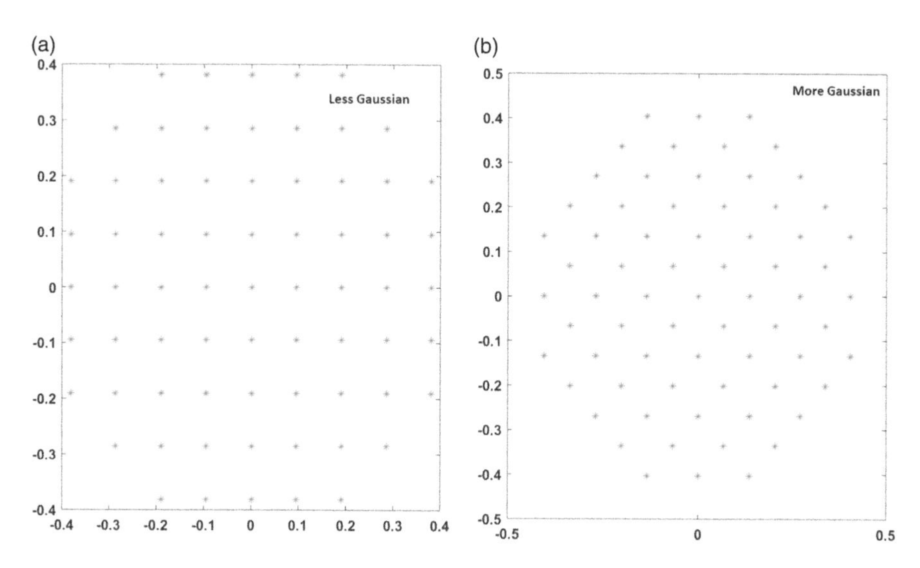

Fig. 5.3 (**a**) Original data with less Gaussian. (**b**) Data with more Gaussian (obtained using PSO)

```
W=rand(2,2);
W(:,1)=W(:,1)/sqrt(sum(W(:,1).^2));
W(:,2)=W(:,2)-(W(:,2)'*W(:,1))*W(:,1);
W(:,2)=W(:,2)/sqrt(sum(W(:,2).^2));
W_init{i}=W;
VAL1=[VAL1 objfun(W,x)];
W=rand(2,2);
W(:,1)=W(:,1)/sqrt(sum(W(:,1).^2));
W(:,2)=W(:,2)-(W(:,2)'*W(:,1))*W(:,1);
W(:,2)=W(:,2)/sqrt(sum(W(:,2).^2));
W_next{i}=W;
VAL2=[VAL2 objfun(W,x)];
end
BESTVAL=[];
BESTSOL=[];
for iter=1:1:10
[P,Q]=min(VAL2);
[P1,Q1]=min(VAL1);
BESTVAL=[BESTVAL VAL1(Q1)];
BESTSOL{iter}=W_init{Q1};
%Global and the Actual movement
VAL3=[];
for i=1:1:2
W=W_init{i}+0.9*(W_next{i}-W_init{i})+...
0.9*(W_next{Q}-W_init{i});
```

```
W(:,1)=W(:,1)/sqrt(sum(W(:,1).^2));
W(:,2)=W(:,2)-(W(:,2)'*W(:,1))*W(:,1);
W(:,2)=W(:,2)/sqrt(sum(W(:,2).^2));
W_move{i}=W;
VAL3=[VAL3 objfun(W,x)];
end

for i=1:1:2
if(VAL3(i)<VAL2(i))
    W_next{i}=W_move{i};
end
W_init{i}=W_move{i};
end
VAL1=[];
VAL2=[];
VAL3=[];
for i=1:1:2
W=W_init{i};
W(:,1)=W(:,1)/sqrt(sum(W(:,1).^2));
W(:,2)=W(:,2)-(W(:,2)'*W(:,1))*W(:,1);
W(:,2)=W(:,2)/sqrt(sum(W(:,2).^2));
W_init{i}=W;
VAL1=[VAL1 objfun(W,x)];
W=W_next{i};
W(:,1)=W(:,1)/sqrt(sum(W(:,1).^2));
W(:,2)=W(:,2)-(W(:,2)'*W(:,1))*W(:,1);
W(:,2)=W(:,2)/sqrt(sum(W(:,2).^2));
W_next{i}=W;
VAL2=[VAL2 objfun(W,x)];
end
end
[U,V]=min(BESTVAL);
pdata=BESTSOL{V}*x;
subplot(1,2,2)
plot(pdata(1,:),pdata(2,:),'r*')
figure
plot(BESTVAL)
```

```
%obj.m
function [res]=objfun(w,x)
res=w*x;
res=kurt(res(1,:))+kurt(res(2,:));
```

```
function [res]=kurt(x)
res=(sum(x.^4)/length(x))-3*(sum(x.^2)/length(x))^2;
res=abs(res);
```

5.2 ANT Colony Technique to Identify the Order in Which the SVM Blocks Are Arranged

ANT colony optimization is based on the inspiration of the ANT's strategy to find the shortest path to reach the destination. When the ants are moving, it secretes the chemical known as pheromone. Based on the higher concentration of the pheromone, the following ants chose the path to move forward. This analogy is adopted to formulate the ANT colony optimization technique to identify the order in which the SVM blocks (refer Fig. 5.4) are arranged. Let λ_{i1} and λ_{i2} be the probability of false-alarm and the probability of miss of the ith block, respectively. The order of the SVM blocks is identified such that the probability of correct decision (given below) is maximized.

$$J = \frac{1}{4}\sum_{j=1}^{j=3} \Pi_{i=1}^{i=j-1}(1-\lambda_{i1})(1-\lambda_{j2}) + \frac{1}{4}\Pi_{i=1}^{i=3}(1-\lambda_{i1}) \tag{5.1}$$

5.2.1 *ANT Colony Optimization (ACO)*

1. The arbitrary order in which the SVM blocks are arranged is an example of the path of the typical ant. Thus each ant (corresponding path) holds the corresponding functional value (J).
2. Initialize the path of the 100 ants and compute the corresponding cost functions.
3. Formulate the pheromone matrix PM (with size $N \times N$) filled up with zeros.
4. For the typical ant's path, the $(i, j)^{\text{th}}$ position of the pheromone matrix PM is said to be active if j^{th} position of the path is having the value i. Update the active positions of the pheromone matrix PM (corresponding to the ith ant's path) with the functional value of the ith ant's path.
5. Obtain the new group of Ant's path using the latest updated pheromone matrix. This is obtained as follows.

   - Note that each ant is described by N elements with each element filled up with distinct number between 1 and N.

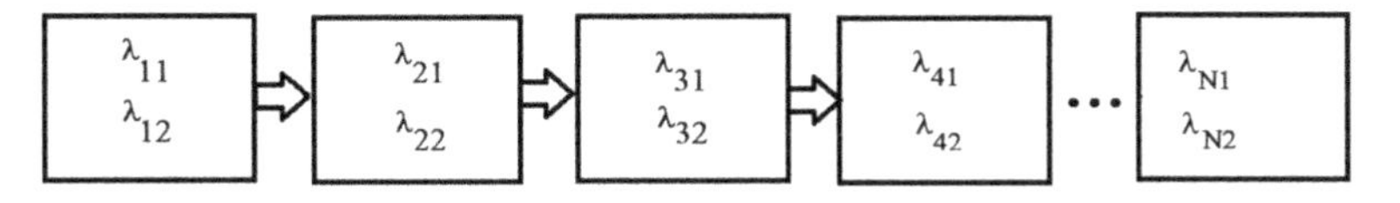

Fig. 5.4 Illustration of cascade of N SVM blocks. λ_{i1} is the probability of miss of the ith block

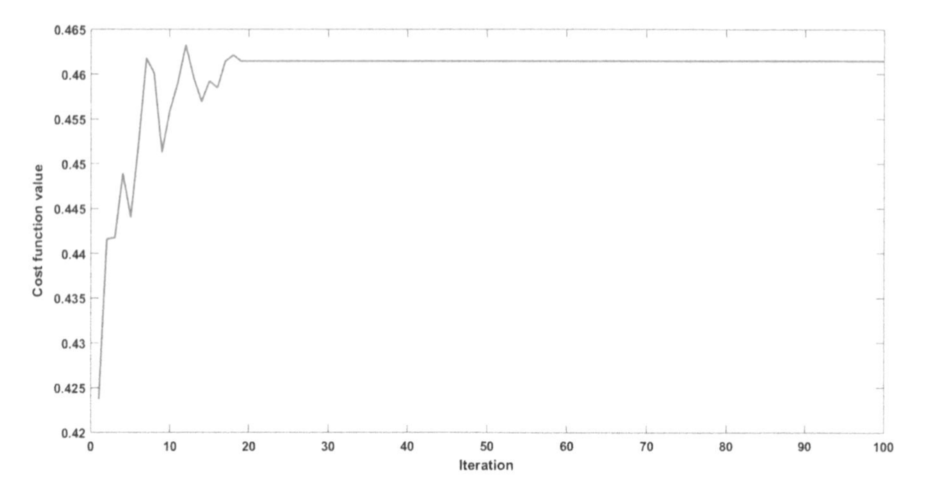

Fig. 5.5 Illustration of maximization of objective function (using ANT colony technique)

- For the typical ant's path, generate the order in which the elements are filled up with. Let it be $p_1, p_2, ..., p_N$.
- In the p_1^{st} column of the Pheromone Matrix PM, identify the index of the row of the maximum value. Let it be q_1. Fill the p_1^{st} element of the typical ant with q_1. This is followed by identifying the index corresponding to the maximum of the p_2^{nd} column. It is chosen as $q_2 \neq q_1$. This is further followed to get the elements of the typical ant path in the next iteration.
- The above steps are repeated 100 times to obtain the path of 100 ants.

6. This is the single iteration.
7. Go to step 4 to start the new iteration.
8. The ant's path corresponding to the largest value in every iteration is collected.
9. Best among the collected paths is declared as the final optimal path (obtained using ACO). The illustrations on maximizing the objective function using ANT colony technique and the attained optimal order are illustrated in Figs. 5.5 and 5.6.

antcolony-prob-detection.m

```
%Montecarlo simulation to determine the order that
maximizes the
%probability of detection
lambda1=rand(1,16)*0.3;
lambda2=rand(1,16)*0.3;
LAMBDA=[lambda1;lambda2];
V=probcorrect(LAMBDA');
%ANTCOLONY STARTS
MATRIX=zeros(16,16);
```

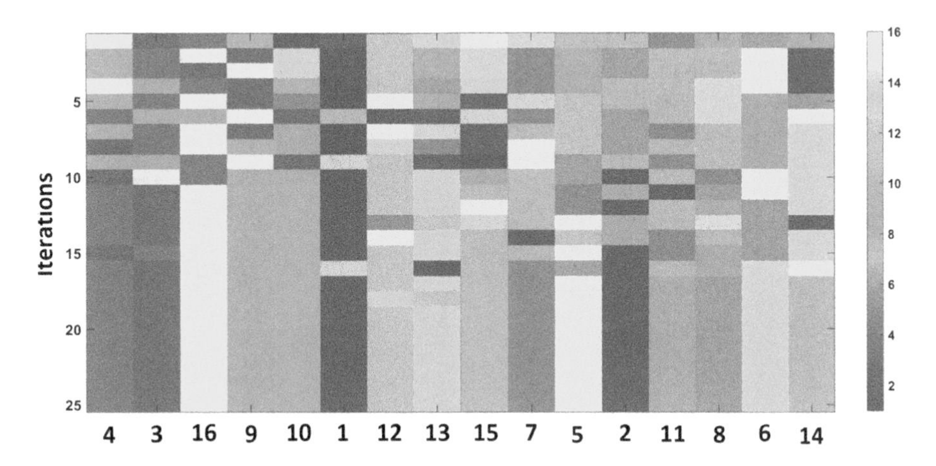

Fig. 5.6 Illustration of obtaining the optimal order (using ANT colony technique)

```
BESTSOL=[];
BESTVAL=[];
N=100;
for i=1:1:N
[P,Q]=sort(rand(1,16));
ORDER{i}=Q;
LAMDAORDER{i}=LAMBDA(:,Q);
end
for iteration=1:1:100
for i=1:1:N
LAMDAORDER{i}=LAMBDA(:,ORDER{i});
VAL(i)=probcorrect(LAMDAORDER{i}');
end
[P2,Q2]=max(VAL);
BESTSOL{iteration}=ORDER{Q2};
BESTVAL(iteration)=P2;
%UPDATAE THE MATRIX
for i=1:1:16
for j=1:1:N
MATRIX(i,ORDER{j}(i))=VAL(j)+0.5*MATRIX(i,ORDER{j}
(i));
end
end
%Selecting the next set of order
temp=zeros(1,16);
COL=[];
for j=1:1:N
    COL=[];
```

```
    [P,Q]=sort(rand(1,16));
    temp1=MATRIX(Q(1),:);
    [P1,Q1]=max(temp1);
    temp(Q(1))=Q1(1);
    COL=[COL Q1];
    for i=2:1:16
    temp1=MATRIX(Q(i),:);
    temp1(COL)=-1000;
    [P1,Q1]=max(temp1);
    temp(Q(i))=Q1(1);
    COL=[COL Q1];
    end
    ORDER{j}=temp;
end
end
[U,V]=max(BESTVAL);
ORDERCHOSEN=BESTSOL{V(1)};
plot(BESTVAL)
xlabel('Iteration')
ylabel('Cost function value')
```

probcorrect.m

```
function [res]=probcorrect(lambda)
%lambda is the vector
N=length(lambda);
s=0;
for j=1:1:N-1
    temp=1;
    for i=1:1:j-1
        temp=temp*(1-lambda(i,1));
    end
    temp=temp*(1-lambda(j,2));
    s=s+temp;
end
temp1=1;
for k=1:1:N-1
    temp1=temp1*(1-lambda(k,1));
end
res=s+temp1;
res=res/N;
```

5.3 Social Emotional Optimization Algorithm (SEOA) for k-Means Clustering

The analogy of SEOA is that the qualities of the human are represented using the vector (solution), and the corresponding functional value is the status of that particular person. Interactions between humans pave the way to change the qualities of the particular human, and hence the status of the human in the society will also get changed. SEOA formulates the procedure that tries to improve the status of the individual human. The steps involved in minimizing the cost function J using SEOA are as follows:

$$J = \sum_{k=1}^{k=K} \sum_{n=1}^{n=N} (x_n t_{nk} - \mu_k t_{nk})^2 \tag{5.2}$$

1. Initialize the quality vectors of n persons and the corresponding status (cost function to be optimized). Let it be $v_1, v_2, ..., v_n$ and the corresponding functional values $f_1, f_2, ..., f_n$. Initialize the Emotion Index of the individual vectors represented as $EI_1, EI_2, ..., EI_n$.
2. Identify L vectors among the n vectors that have the maximum cost (low status) functional values. Let it be $u_1, u_2, ..., u_L$.
3. Introduce the perturbations in the quality of the vectors using the obtained L vectors to obtain the new sets of vectors v_i' for $i = 1 \cdots n$, as

$$v_i' = v_i - \alpha \times rand \sum_{j=1}^{j=L} (\mathbf{v_i} - \mathbf{u_j}) \tag{5.3}$$

4. Let the functional values associated with v_i' is computed as f_i'.
5. If $((f_i < f_i'))$, then $EI_i = EI_i - \beta$.
6. Compute $k = arg_r min_{i=1}^{n} f_i$ as the Global vector v_g.
7. If $((f_i < f_i'))$, $v_{local}^i = v_i$, else $v_{local}^i = v_i'$.
8. Update the vectors based on the values of the individual Emotion Index (EI) as follows:
   Case 1: If $(EI(i) < Lower\ threshold)$

$$f_i'' = f_i' + \alpha \times rand(\mathbf{v_i} - v_g) \tag{5.4}$$

   Case 2: If $(Lower\ threshold < EI(i) < Upper\ threshold)$

$$f_i'' = f_i' + \alpha \times rand(\mathbf{v_i} - v_g) - \alpha \times rand \sum_{j=1}^{j=L} (\mathbf{v_i} - \mathbf{u_j}) \tag{5.5}$$

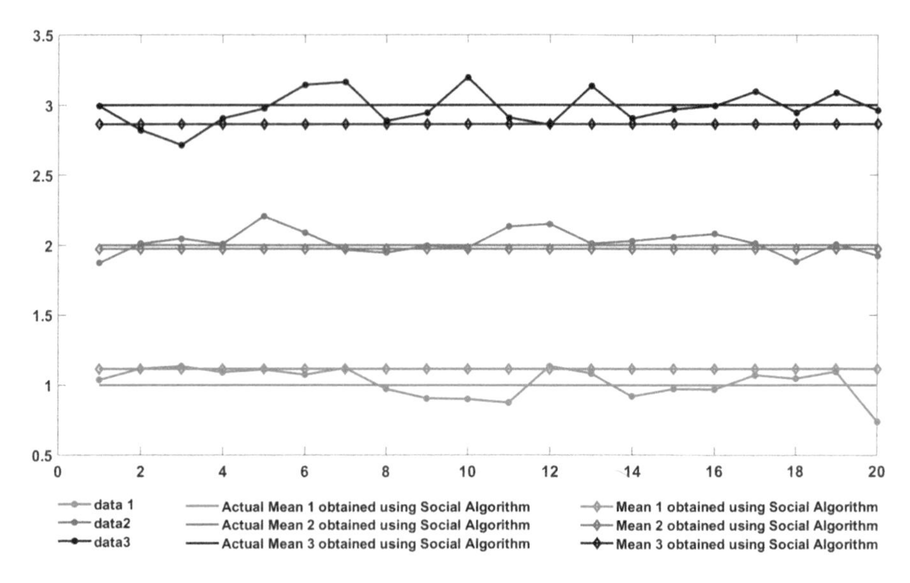

Fig. 5.7 Optimization results obtained using SEOA by minimizing J

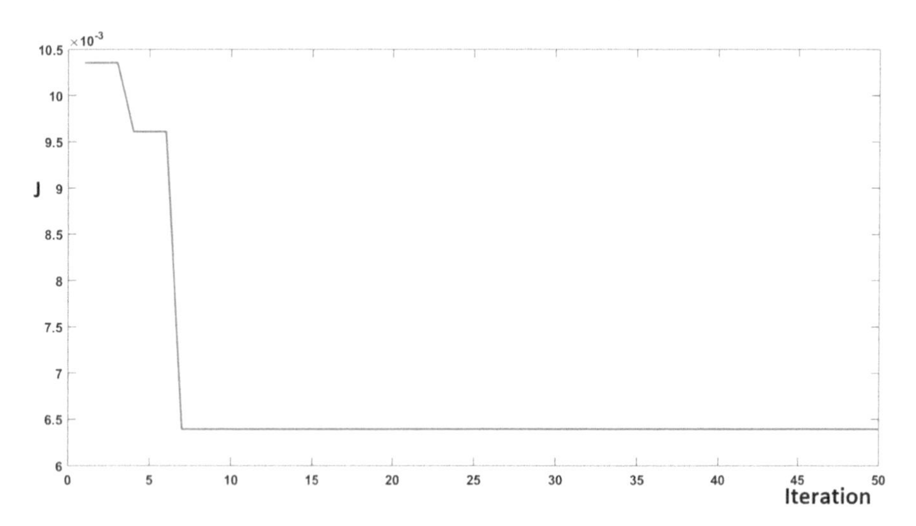

Fig. 5.8 Illustration of minimizing J using SEOA

Case 3: If $(EI(i) > Upper\ threshold)$

$$f_i'' = f_i' + \alpha \times rand(\mathbf{v_i} - v_{local}^i) \tag{5.6}$$

The illustration of the result obtained using SEOA by minimizing J and corresponding convergence graph is given in Figs. 5.7 and 5.8.

seoakmeans.m

```
X1=randn(1,20)*0.1+1;
X2=randn(1,20)*0.1+2;
X3=randn(1,20)*0.1+3;
X=[X1 X2 X3];
EI=ones(1,200);
%EI indicates Emotion Index
%Initialize population
for i=1:1:200
M{i}=rand(1,3)*5;
end
BESTCOL=[];
BESTVAL=[];

VAL=[]
for i=1:1:200
VAL=[VAL socialalgo(X,M{i})];
end
[P,Q]=sort(VAL);
L=3;
GLOBALBEST=M{Q(1)};
GLOBALBESTVAL=P(1);
%Obtain the next population.
for i=1:1:200
    M1{i}=M{i}-(1/10)*rand*(M{i}-M{Q(100)})- ...
(1/10)*rand*(M{i}-M{Q(99)})-(1/10)*rand*(M{i}-
M{Q(98)})
end

for i=1:1:200
    if (socialalgo(X,M1{i})<socialalgo(X,M{i}))
    LOCALBEST{i}=M1{i};
    else
    LOCALBEST{i}=M{i};
    end
end


for iteration=1:1:50
for i=1:1:200
    if(socialalgo(X,M1{i})>socialalgo(X,M{i}))
        EI(i)=EI(i)-0.01;
    end
if(socialalgo(X,M1{i})<GLOBALBESTVAL)
    EI(i)=1;
```

```
end
end


VAL=[]
for i=1:1:200
VAL=[VAL socialalgo(X,M1{i})];
end
[P,Q]=sort(VAL);
if(VAL(Q(1))<socialalgo(X,GLOBALBEST))
    GLOBALBEST=M1{Q(1)};
    GLOBALBESTVAL=VAL(Q(1));
end

for i=1:1:200
    if (socialalgo(X,M1{i})<LOCALBEST{i})
    LOCALBEST{i}=M1{i}
    end
end
for i=1:1:200
        if(EI(i)<0.25)
        M2{i}=M1{i}+(1/10)*rand*(M1{i}-GLOBALBEST);
        elseif (0.25<EI(i)<0.75)
        M2{i}=M1{i}+(1/10)*rand*(M1{i}-...
GLOBALBEST)+(1/10)*rand*(M1{i}-LOCALBEST{i})...
        -(1/10)*rand*(M1{i}-M1{Q(100)})...
        -(1/10)*rand*(M1{i}-M1{Q(99)})...
        -(1/10)*rand*(M1{i}-M1{Q(98)});
        else
        M2{i}=M1{i}+(1/10)*rand*(M1{i}-LOCALBEST{i});
        end
        end
M=M1;
M1=M2;
BESTVAL=[BESTVAL GLOBALBESTVAL];
BESTSOL{iteration}=GLOBALBEST;
end
FINALSOL=BESTSOL{50};
G=[];
for i=1:20:60
    [p,q]=min((repmat(X(i),1,length(FINALSOL))-
    FINALSOL).^2);
    G=[G q];
end
figure
```

```
plot(X1,'r*-')
hold on
plot(X2,'b*-')
plot(X3,'k*-')
plot(ones(1,length(X1))*FINALSOL(G(1)),'rd-')
plot(ones(1,length(X2))*FINALSOL(G(2)),'bd-')
plot(ones(1,length(X3))*FINALSOL(G(3)),'kd-')
plot(ones(1,length(X1))*1,'r-')
plot(ones(1,length(X2))*2,'b-')
plot(ones(1,length(X3))*3,'k-')
```

socialalgo.m

```
function [s]=socialalgo(X,M)
d=0;
t=zeros(length(X),length(M));
for i=1:1:length(X)
    [p,q]=min((repmat(X(i),1,length(M))-M).^2);
    t(i,q)=1;
end
s=0;
for i=1:1:length(X)
    for j=1:1:length(M)
    s=s+(X(i)*t(i,j)-M(j)*t(i,j))^2 ;
    end
end
s=s/(length(M)*length(X));
```

5.4 Social Evolutionary Learning Algorithm (SELA) to Optimize the Tuning Parameter σ for KLDA

The analogy for SELA algorithm is as follows: the vector associated with the individual members of the corresponding families describes the learned material, and the corresponding cost functional value associated with the individual vector is the IQ level of the corresponding member. The algorithm involves interactions between the members of the within family and between family members to improvise the IQ level of the individual members as described as follows:

1. Consider 40 families. Each family is having two parents and two kids. The feature describing the 40 members is represented. $F_i^{P_j}$ and $F_i^{K_j}$ is the jth parent member and j^{th} kid of the ith family, respectively. Also the corresponding functional values are represented as $V_i^{P_j}$ and $V_i^{K_j}$, respectively.

2. The outcome of the variable $F_i^{P_j}$ is randomly chosen between $FL_i^{P_j}$ and $FU_i^{P_j}$. Similarly, the outcome of the variable $F_i^{K_j}$ is randomly chosen between $FL_i^{K_j}$ and $FU_i^{K_j}$,
3. Every iteration involves in adjusting the ranges associated with the individual member.
4. Initialize the ranges associated with the individual members of 40 families.
5. Sample the outcome for the individual members and compute the functional values associated with the corresponding members.
6. Identify the member that has the least cost function. Let it be the mth parent from the nth family or m^{th} kid from the nth family.
7. Adjust the range of the individual members for parents of all the family (except the family member m^{th} parent from the nth family) using interaction with arbitrarily chosen other parent (say r^{th} parent of sth family) from arbitrary family as follows:

$$FL_i^{p_j} = FL_i^{p_j} - k \times rand \times ||FL_s^{p_r}||_2 \tag{5.7}$$

$$FH_i^{p_j} = FH_i^{p_j} + k \times rand \times ||FL_s^{p_r}||_2 \tag{5.8}$$

8. For the mth parent from the nth family (if it has the least functional value), the update is as follows:

$$FL_n^{p_m} = FL_n^{p_m} - k \times rand \times ||FL_n^{p_m}||_2 \tag{5.9}$$

$$FH_n^{p_m} = FH_n^{p_m} + k \times rand \times ||FL_n^{p_m}||_2 \tag{5.10}$$

9. For the kid membership (except for the mth kid from the nth family (if it has the least function)), the adjustment is done by choosing the kids of arbitrarily chosen (more chances are given to the kid member with least cost) other family (with probability $pk1$) or parents of other family (more chances are given to the parent member with least cost) (with probability $(1 - pk1)$).
10. For the mth kid from the nth family (if it has the least function), the update is done by choosing the parents (with probability $pk2$) (equal chance among the parents of the nth family) or kids of the nth family (equal chance among the kids of the nth family).
11. Best member in each iterations is tracked, and best member in the last iteration is declared as the solution to the optimization problem.

The illustration on data achieved using SELA and the corresponding convergence graph are given in Figs. 5.9 and 5.10, respectively.

SELO-sigma-KLDA.m

```
%SELO algorithm
%Consider 10 families with 2 parents
%and 2 kids in each family is
```

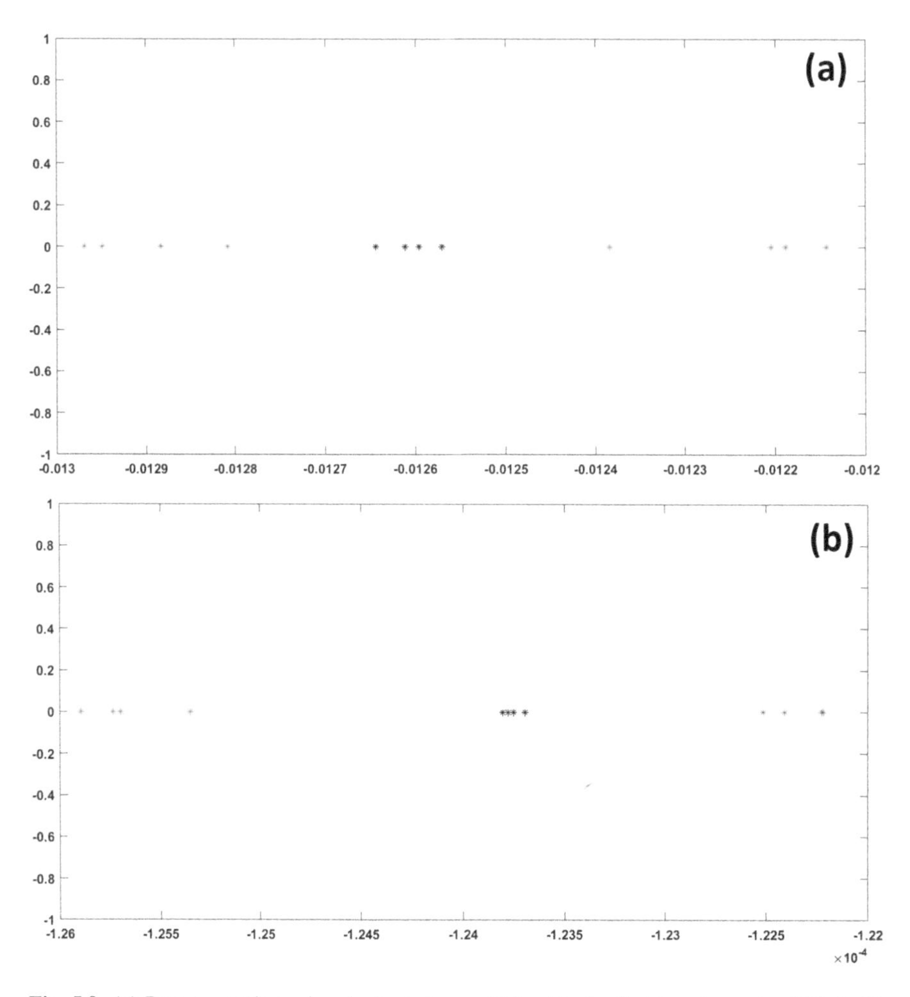

Fig. 5.9 (**a**) Data separation using the best sigma obtained in the first iteration using SELA. (**b**) Data separation using the best sigma obtained in the last iteration using SELA

```
%considered.
%Training vectors subjected to KLDA.
pk1=0.3;
pk2=0.7;
X1=randn(3,4)*10+ones(3,4)*1;
X2=randn(3,4)*10+ones(3,4)*2;
X3=randn(3,4)*10+ones(3,4)*3;
%Construction of Gram-matrix
M=[X1 X2 X3];
%Initialize population (with 10 families,
%2 parents and 2 kids in each family)
```

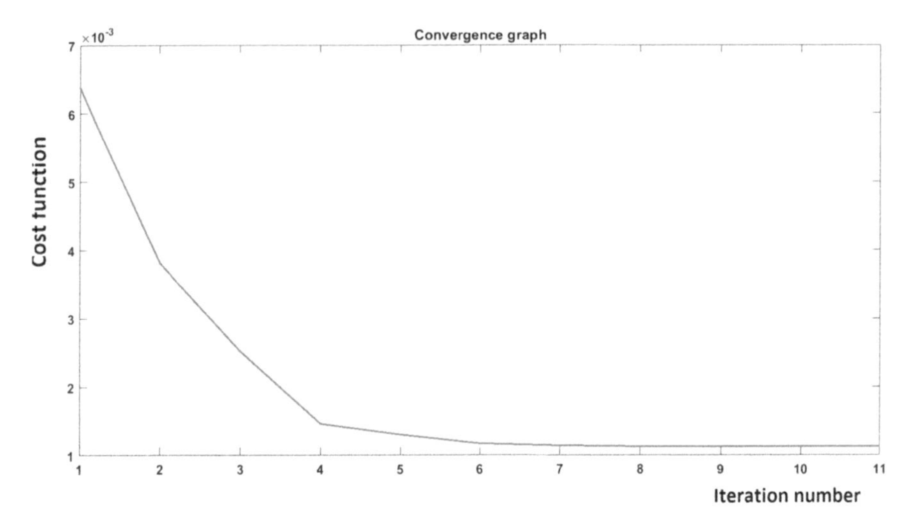

Fig. 5.10 Convergence obtained using SELA

```
%Initially the ranges are identical for two parents
%and two kids
FL=[0.1 1.5 2.1 3.1 4.1 5.1 10 20 30 40];
FU=[1 2 3 4 5 10 20 30 40 50];
FL=[FL FL FL FL];
FU=[FU FU FU FU];
BESTVAL=[];
BESTSOL=[];
V=[];
F=[];
for i=1:1:40
F(i)=0.1*rand*(FU(i)-FL(i))+FL(i);
[PD,value]=seloalgo(M,F(i));
V=[V value];
end
%Identifying the best member
[P,Q]=min(V);
[PD,value]=seloalgo(M,F(Q));
BESTVAL=value;
BESTSOL=F(Q);
figure
plot(PD(1:1:4),zeros(1,4),'r*')
hold on
plot(PD(5:1:8),zeros(1,4),'b*')
plot(PD(9:1:12),zeros(1,4),'k*')
```

```
for iteration=1:1:20
V=[];
F=[];
for i=1:1:40
F(i)=FL(i)+0.1*rand*(FU(i)-FL(i));
[PD,value]=seloalgo(M,F(i));
V=[V value];
end
%Identifying the best member
[P,Q]=min(V);
if(V(Q)<BESTVAL(length(BESTVAL)))
    [PD,value]=seloalgo(M,F(Q));
    BESTVAL=[BESTVAL value];
    BESTSOL=[BESTSOL F(Q)];
end

%Adjusting the range for the parents
for i=1:1:20
if(mod(Q,10)~=i)
temp=1./V(1:1:20);
R=cumsum(temp);
[temp]=cumsum(R)-rand;
[g,h]=find(temp>0);
FL(i)=F(i)-0.1*rand*norm([FL(h) FU(h)]);
FU(i)=F(i)+0.1*rand*norm([FL(h) FU(h)]);
else
if(Q<=20)
FL(Q)=F(Q)-0.1*rand*norm([FL(Q) FU(Q)]);
FU(Q)=F(Q)+0.1*rand*norm([FL(Q) FU(Q)]);
end
end
end

%%Update for kids
for i=1:1:20
if(mod(Q,10)~=i)
 if(rand>pk1)
temp=1./V(21:1:40);
R=cumsum(temp);
[temp]=cumsum(R)-rand;
[g,h]=find(temp>0);
FL(20+i)=F(i+20)-0.1*rand*norm([FL(h+20) FU(h+20)]);
FU(20+i)=F(i+20)+0.1*rand*norm([FL(h+20) FU(h+20)]);
else
temp=1./V(1:1:20);
```

```
R=cumsum(temp);
[temp]=cumsum(R)-rand;
[g,h]=find(temp>0);
FL(20+i)=F(i+20)-0.1*rand*norm([FL(h) FU(h)]);
FU(20+i)=F(i+20)+0.1*rand*norm([FL(h) FU(h)]);
end

else
if(Q>20)
if(rand>pk2)
if(rand>0.5)
FL(Q)=F(Q)-0.1*rand*norm([FL(i+20) FU(i+20)]);
FU(Q)=F(Q)+0.1*rand*norm([FL(i+20) FU(i+20)]);
else
FL(Q)=F(Q)-0.1*rand*norm([FL(i+30) FU(i+30)]);
FU(Q)=F(Q)+0.1*rand*norm([FL(i+30) FU(i+30)]);
end
else
if(rand>0.5)
FL(Q)=F(Q)-0.1*rand*norm([FL(i) FU(i)]);
FU(Q)=F(Q)+0.1*rand*norm([FL(i) FU(i)]);
else
FL(Q)=F(Q)-0.1*rand*norm([FL(i+10) FU(i+10)]);
FU(Q)=F(Q)+0.1*rand*norm([FL(i+10) FU(i+10)]);
end
end
end
end
end
end
[PD,value]=seloalgo(M,BESTSOL(length(BESTSOL)));
figure
plot(PD(1:1:4),zeros(1,4),'r*')
hold on
plot(PD(5:1:8),zeros(1,4),'b*')
plot(PD(9:1:12),zeros(1,4),'k*')
figure
plot(BESTVAL)
title('Convergence graph')
seloalgo.m
function [PD,res]=seloalgo(M,sigma)
%Tuning the sigma value for KLDA to
%maximize the objective function J=trace(E'*SB*E)/
    trace(E'*SW*E)
%(Computed using kernel function)
```

```
for i=1:1:12
     for j=1:1:12
G(i,j)=gaussiankernel(M(:,i),M(:,j),sigma);
     end
end
%Between-class scatter matrix
C1=mean(G(:,1:1:4)')';
C2=mean(G(:,5:1:8)')';
C3=mean(G(:,9:1:12)')';
C=(C1+C2+C3)*(1/3);
SB=(C1-C)*(C1-C)'+(C2-C)*(C2-C)'+(C3-C)*(C3-C)';
SB=4*SB/(4-3);
SW=(G(:,1:1:4)-repmat(C1,1,4))*(G(:,1:1:4)-repmat
        (C1,1,4))'+...
     (G(:,5:1:8)-repmat(C2,1,4))*(G(:,5:1:8)-repmat
        (C2,1,4))'+...
     (G(:,9:1:12)-repmat(C3,1,4))*(G(:,9:1:12)-repmat
        (C3,1,4))';
SW=SW/(12-3);
[E,D]=eig(pinv(SW)*SB);
E=real(E(:,1));
res=abs(trace(E'*SW*E)/trace(E'*SB*E));
res=real(res);
PD=E'*G;
```

5.5 Genetic Algorithm to Obtain the Equation of the Line That Partitions Two Classes in 2D Space

This is inspired from the genetic evolution of the living organisms. The chromosomes in the genetic algorithm form the search space, and the cost function associated with the chromosomes is the ability to perform cross over with other chromosomes. The algorithm is as follows:

1. Initialize n number of chromosomes $(C_1, C_2, ..., C_n)$ and compute the cost associated with the individual chromosomes where the corresponding functional values are $f_1, f_2, ..., f_n$.
2. Two chromosomes are randomly chosen (say m and n) and subject to cross over to obtain the two chromosomes as follows:

$$C_{new1} = \alpha \times C_m + \beta \times C_n \tag{5.11}$$

$$C_{new2} = \beta \times C_m + \alpha \times C_n \tag{5.12}$$

3. Repeat step 2 n times to obtain $2n$ chromosomes

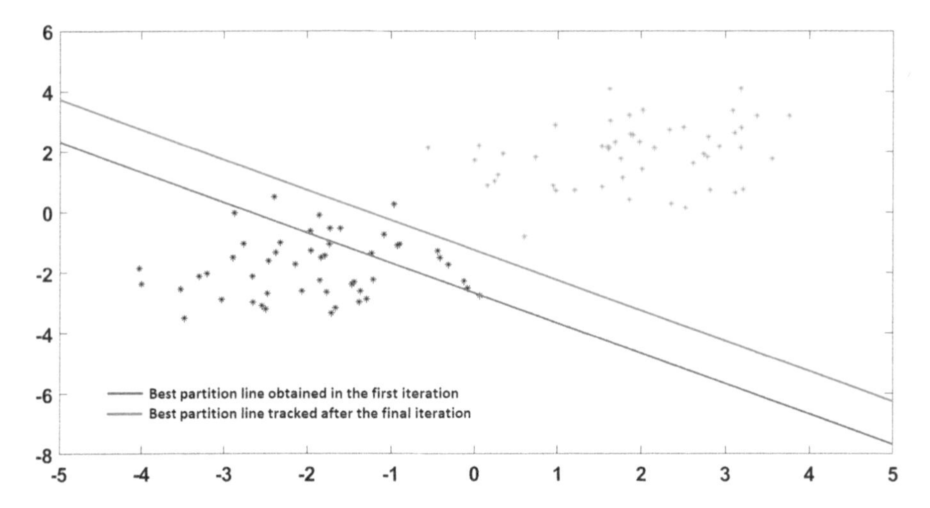

Fig. 5.11 Optimization results (equation of the line) obtained using genetic algorithm by maximizing J_1

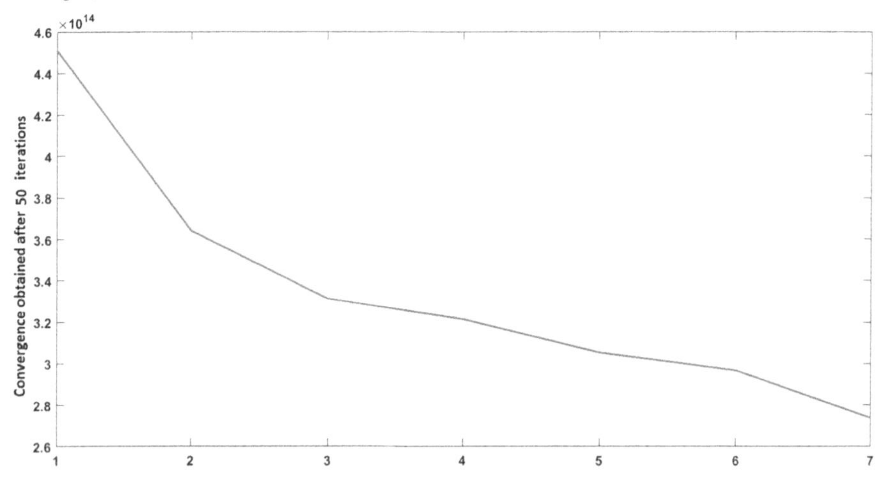

Fig. 5.12 Convergence obtained using genetic algorithm by maximizing J_1

4. Select $m << 2n$ chromosomes randomly and subject to mutations as follows: $C_{after\ mutation} = rand \times C_{before\ mutation}$
5. Select n chromosomes from the generated chromosomes (with more chances given to the one with higher cost functional value). This is one iteration.
6. Repeat 2–5 for finite number of iterations.
7. The best in every iteration is tracked, and the best among the best is declared as the optimized value.
8. The illustration on determining the equation of the line $w_1x + w_2y + b$ described by the vector $[w_1\ w_2\ b]$ is optimized using genetic algorithm.

The illustration of genetic algorithm is given in Figs. 5.11 and 5.12.

GApartition.m

```
clear all
close all
%Identifying the line that
%divides the two class as follows.
x=randn(2,50)+2;
y=randn(2,50)-2;
z=[x y];
plot(x(1,:),x(2,:),'r*')
hold on
plot(y(1,:),y(2,:),'k*')
t=[ones(1,50) ones(1,50)*(-1)];
%Generate data
IV=[]
for i=1:1:100
w{i}=rand(1,2)-1;
b(i)=rand*2-1;
IV=[IV fga(w{i},z,b(i),t')];
end
[A,B]=max(IV);
BESTVAL=[A];
BESTSOL=[w{B} b(B)];
index=1;
for iteration=1:1:50
F=[];
%Cross-over
for i=1:2:200
    r1=round(rand*24)+1;
    r2=round(rand*24)+1;
    r=rand;
    DATAW{i}=r*w{r1}+(1-r)*w{r2};
    DATAW{i+1}=(1-r)*w{r1}+r*w{r2};
    DATAb(i)=r*b(r1)+(1-r)*b(r2);
    DATAb(i+1)=(1-r)*b(r1)+r*b(r2);
    F=[F fga(DATAW{i},z,DATAb(i),t') fga(DATAW{i+1},
    z,DATAb(i+1),t')];
end
%Selection from the generated data
%Formulation of Roulette wheel
F=F-min(min(F));
F=F/(sum(F)+eps);
F=cumsum(F);
VAL=[];
for i=1:1:100
```

```
    [P,Q]=find((F-rand)>0);
    if(isempty(Q)==1)
    Q=1;
    end
    w{i}=[DATAW{Q(1)}];
    b(i)=DATAb(Q(1));
    VAL=[VAL fga(w{i},z,b(i),t')];
end
[M,N]=max(VAL);
if(BESTVAL(index)<M)
index=index+1;
BESTVAL=[BESTVAL M];
BESTSOL=[BESTSOL;w{N} b(N)];
end
end
INITIALSOL=BESTSOL(1,:);
w=INITIALSOL(1:2);
b=INITIALSOL(3);
x1=-5:0.1:5;
y1=(ones(1,length(x1))-w(1)*x1-ones(1,length(x1))*
b(1))/w(1);
hold on, plot(x1,y1,'b-')

FINALSOL=BESTSOL(size(BESTSOL,1),:);
w=FINALSOL(1:2);
b=FINALSOL(3);
x1=-5:0.1:5;
y1=(ones(1,length(x1))-w(1)*x1-ones(1,length(x1))*
b(1))/w(1);
hold on
plot(x1,y1,'r-')
figure
plot(1./BESTVAL)
fga.m
function [res]=fga(w,x,b,t)
b=ones(1,size(x,2))*b;
res=sum(sum((w*x+b).*t))/length(x);
flag=isnan(res);
if(flag==1)
     res=1;
end
```

5.6 Artificial Neural Network (ANN)

Artificial Neural Network is a biologically inspired algorithm, mimicking the working of the brain. The most often used mathematical model of the brain is the back propagation feed-forward neural network. It consists of layered architecture. Each layer consists of neurons. Every neuron in the input layer is connected to every neuron in the hidden layer. Furthermore, every neuron in the hidden layer is connected to every neuron in the output layer. This is used for (a) regression model, (b) two-class classification model, and (c) multi-class classification model. Let us consider the network with p neurons in the input layer, q neurons in the hidden layer, and r neurons in the output layer.

5.6.1 *Multi-Class Model*

1. Let the input vector be represented as **x**, hidden vector be represented as **h**, and the output vector be represented as **y**.
2. The input vector is appended with 1 to obtain the modified input vector $\mathbf{x_1}$ of size $(p + 1)$.
3. The weight matrix connecting input layer and hidden layer is represented as W^1 (size $(p+1) \times (q)$). In the same fashion, the hidden neuron vector **h** is appended 1 to obtain the modified hidden layer vector **h** with size $(p + 1)$. The weight matrix connecting hidden layer and output layer is represented as W^2 (size $(q + 1) \times r$.)
4. The relationship between the vectors **x**, **h**, and **o** is summarized as follows:

$$\mathbf{h}^T = f_h(\mathbf{x_1} W^1) \tag{5.13}$$

$$\mathbf{y}^T = f_y(\mathbf{h_1} W^2) \tag{5.14}$$

where $f(.)$ is the activation function. Let $\mathbf{a^T} = \mathbf{x_1} W^1$ and $\mathbf{b^T} = \mathbf{h_1} W^2$. In the case of multi-class problem, the activation function in the hidden layer f_h is chosen as the sigmoidal or tanh function as follows:

$$y_i = f_h(a_i) = \frac{1}{1 + e^{-a(i)}}$$

or

$$y_i = f_h(a_i) = \frac{e^{-a(i)} - e^{a(i)}}{e^{-a(i)} + e^{-a(i)}}$$

Also, the activation function in the output layer f_y is chosen as the soft-max function as follows:

$$y_i = \frac{e^{-b(i)}}{\sum_{j=1}^{j=r} e^{-b(j)}} \tag{5.15}$$

The training data consists of set of input vectors **x** and the corresponding target vectors **t**. The weight matrix is optimized by minimizing the cross-entropy function as follows:

$$-\sum_{n=1}^{n=N} t_n log(y_n) \tag{5.16}$$

where N is the number of training data used to train the network.

The steps involved in updating the weights are summarized as follows:

1. Initialize weights W^1 (size: $(p+1) \times (q)$) and W^2 (size $((q+1) \times r)$).
2. Update the weight matrices using sequential learning.
3. Compute the error vector (gradients in the output layer), e_1 and $e_2 \cdots e_r$, corresponding to r neurons in the output layer corresponding to the nth input vector.
4. The updated equation for the weight matrices is given as follows:

$$W^2(t+1) = W^2(t) - \eta \mathbf{h_1} \mathbf{error} \tag{5.17}$$

The ith element of the row vector **error** is e_i.

5.

$$\text{Let } temp_r^{1:q} = e_r(W_2^r)^T \cdot (\mathbf{h_1})^T \cdot (1-(\mathbf{h_1})^T) \tag{5.18}$$

$$W^1(t+1) = W^1(t) - \eta x_1^T \cdot (temp_1^{1:q} + \cdots + temp_r^{1:q}) \tag{5.19}$$

where $\cdot$ is the point operator, W_r^2 is the rth column vector of the matrix W^2, $\mathbf{x_1^T}$ is the nth input vector arranged row-wise, and $temp_r^{1:q}$ is the first q elements of the vector $temp_r$.

The experiments are performed using the data given in Fig. 5.13. The class-conditional pdf obtained using the trained BPNN and the convergence graph are illustrated in Figs. 5.14 and 5.15. In this case, each neuron in the output indicates class conditional pdf for the individual class and forms the probabilities associated with the partition set.

BPNNtwoclass-softmax.m

```
%Formulating BPNN for two-class problem with
%soft-max activation function Neural Network

temp1=cell2mat(CLASS1');
temp2=cell2mat(CLASS2');
TRAINDATA=[temp1(1:1:5,:) ones(5,1);temp2(1:1:5,:)
ones(5,1)];
TESTDATA=[temp1(6:1:10,:) ones(5,1);temp2(6:1:10,:)
ones(5,1)];
```

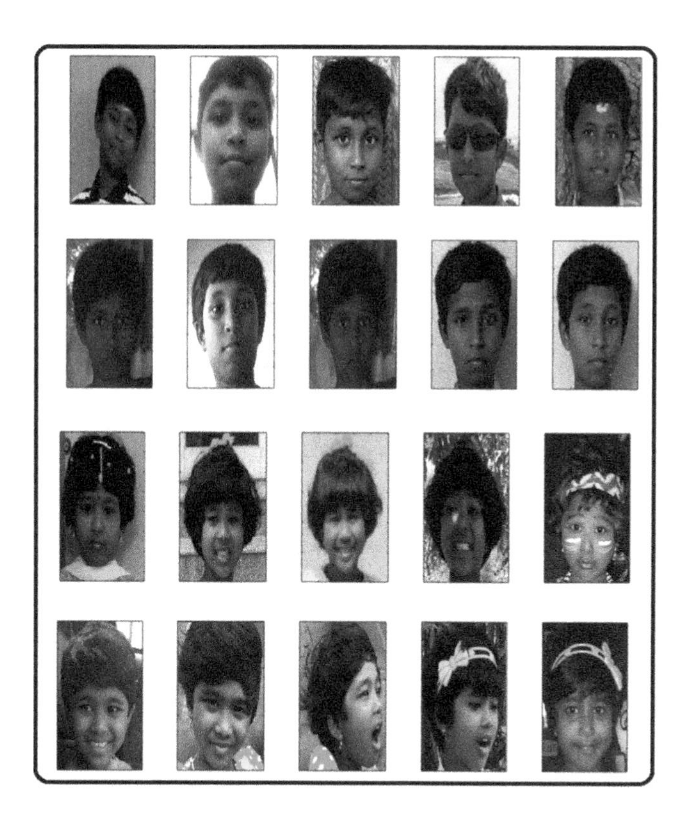

Fig. 5.13 Sample image data subjected to BPNN classifier. First and third rows are the training set for the class 1. Second and fourth rows are the testing data for the class 2

```
TARGET=[ones(1,5) zeros(1,5);zeros(1,5) ones(1,5)];
W1=randn(2501,1000);
W2=randn(2,1001);
JCOL=[];
for iteration=1:1:10
   YCOL=[];
    for n=1:1:10
H=logsig(TRAINDATA(n,:)*W1);
H=[H 1];
Ytemp=H*W2';
%Using soft-max
Y(1)=exp(-Ytemp(1))/(exp(-Ytemp(1))+exp(-Ytemp(2)));
Y(2)=1-Y(1);
YCOL=[YCOL;Y(1) Y(2)];
%Updating weights W2
W2(1,:)=W2(1,:)-0.01*(TARGET(1,n)-Y(1)).*H;
W2(2,:)=W2(2,:)-0.01*(TARGET(2,n)-Y(2)).*H;
%Updating weights W1
```

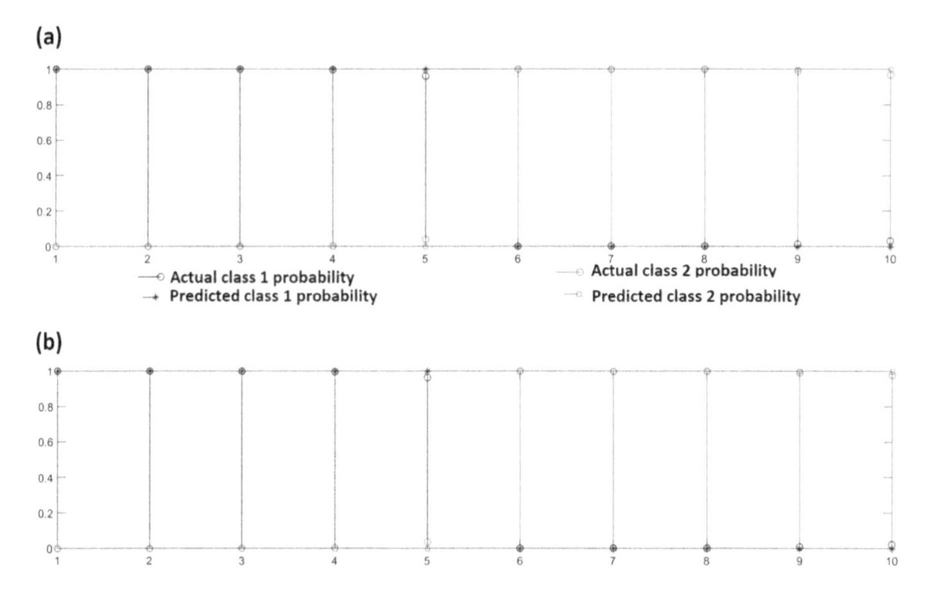

Fig. 5.14 Class probability obtained using the trained BPNN using soft-max objective function. (**a**) Training data. (**b**) Testing data

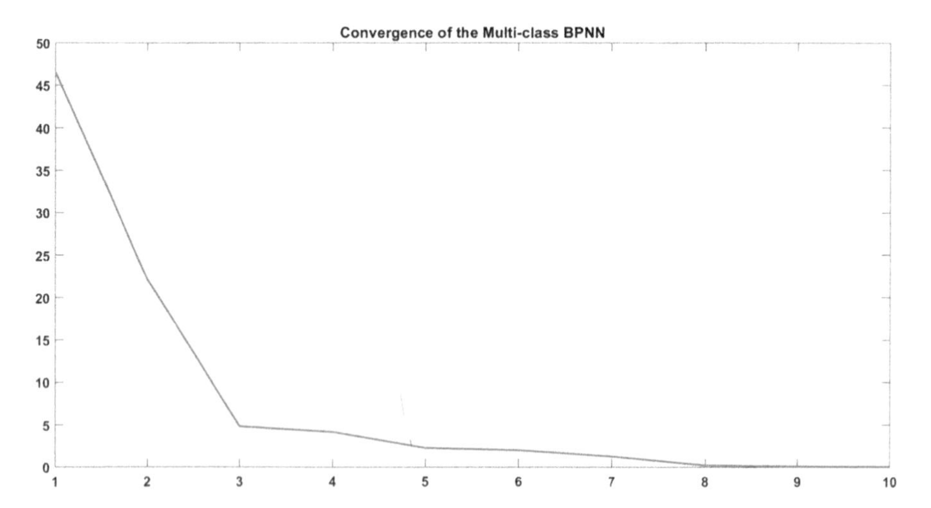

Fig. 5.15 Convergence graph achieved for training BPNN using multi-class entropy as the objective function

```
G1=(TARGET(1,n)-Y(1)).*W2(1,:).*H.*(1-H);
G1=G1(1:1:1000);
G2=(TARGET(2,n)-Y(2)).*W2(2,:).*H.*(1-H);
G2=G2(1:1:1000);
W1=W1-0.01*TRAINDATA(n,:)'*(G1+G2);
end
```

```
JCOL=[JCOL -1*sum(sum(TARGET.*log(YCOL')))];
end
YCOLTRAIN=YCOL;
figure
plot(JCOL)
title('Convergence of the Multi-class BPNN')
YCOLTEST=[];
for n=1:1:10
H=logsig(TESTDATA(n,:)*W1);
H=[H 1];
Ytemp=H*W2';
%Using soft-max
Y(1)=exp(-Ytemp(1))/(exp(-Ytemp(1))+exp(-Ytemp(2)));
Y(2)=1-Y(1);
YCOLTEST=[YCOLTEST;Y(1) Y(2)];
end
figure
subplot(2,1,1)
stem(YCOLTRAIN(:,1),'k-')
hold on
stem(YCOLTRAIN(:,2),'ro')
hold on
stem(TARGET(1,:),'k*')
hold on
stem(TARGET(2,:),'rs')

subplot(2,1,2)
stem(YCOLTEST(:,1),'k-')
hold on
stem(YCOLTEST(:,2),'ro')
hold on
stem(TARGET(1,:),'k*')
hold on
stem(TARGET(2,:),'rs')
```

5.6.1.1 Multiple Two-Class Problem

In the case of multiple two-class problem, every neuron in the output layer is treated as the individual two-class problem. Hence the activation function used in the output layer is the sigmoidal function. Also the objective function is the product of the likelihood function of the two-class cross-entropy computed for the individual neuron in the output layer. Consider the BPNN ($p \times q \times r$), the activation of the rth neuron in the output layer is given as

$$y_i = \frac{1}{1 + e^{-b_i}} \tag{5.20}$$

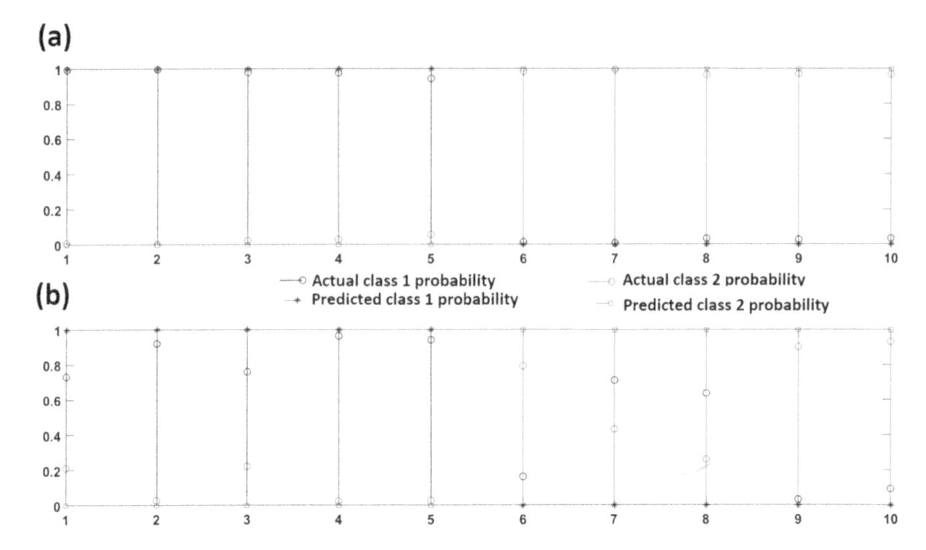

Fig. 5.16 Class probability obtained using the trained BPNN using sigmoidal activation function. (**a**) Training data. (**b**) Testing data

The objective function is given as follows:

$$\Pi_{n=1}^{N}\Pi_{i=1}^{i=r} y_{ni}^{t_{ni}}(1-y_{ni})^{1-t_{ni}} \tag{5.21}$$

Taking negative logarithm, we get the following:

$$-\sum_{n=1}^{N}\sum_{i=1}^{i=r}(t_{ni}log(y_{ni})+(1-t_{ni})log(1-y_{ni}) \tag{5.22}$$

The gradient adjustment (weight update) follows (5.17)–(5.19) with y_i computed using (5.20). Two neurons are used in the output layer. In multi-class (soft-max setup), the two neurons combined together form probabilities associated with the partition set. But in this case, the output of the first neuron indicates the class conditional pdf belonging to class 1 (with the other class as the one not belonging to class 1 (which is not the class 2)). Similarly, the second neuron indicates the class conditional pdf belonging to class 2 (with the other class as the one not belonging to class 2). The class conditional pdf obtained and the corresponding convergence graph using trained BPNN are illustrated in Figs. 5.16 and 5.17.

BPNN-multipletwoclass.m

```
%ANNdemo for two-class problem
load IMAGEDATA
%Formulating Nerual Network
temp1=cell2mat(CLASS1');
temp2=cell2mat(CLASS2');
```

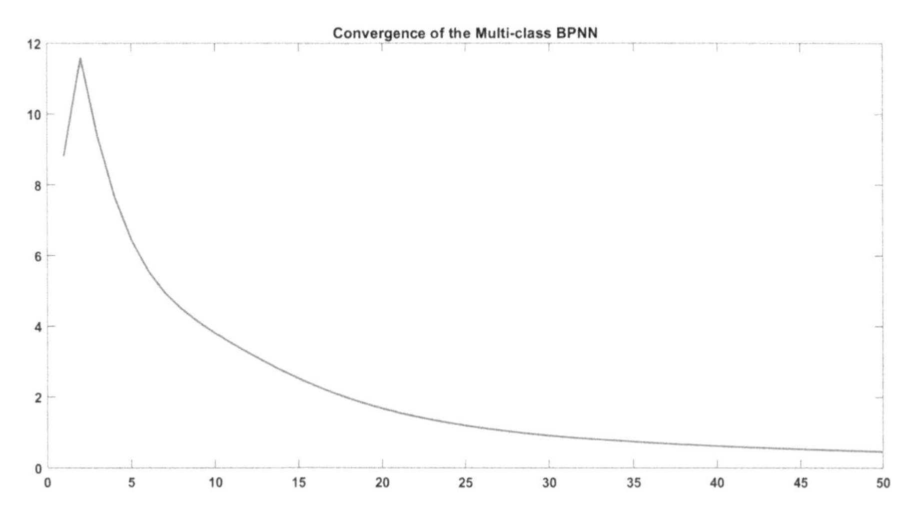

Fig. 5.17 Convergence graph achieved for training BPNN using summation of the two-class entropy as the objective function

```
TRAINDATA=[temp1(1:1:5,:) ones(5,1);temp2(1:1:5,:)
ones(5,1)];
TESTDATA=[temp1(6:1:10,:) ones(5,1);temp2(6:1:10,:)
ones(5,1)];
TARGET=[ones(1,5) zeros(1,5);zeros(1,5) ones(1,5)];
W1=randn(2501,1000)*0.1;
W2=randn(2,1001)*0.1;
JCOL=[];
for iteration=1:1:50
   YCOL=[];
for n=1:1:10
H=logsig(TRAINDATA(n,:)*W1);
H=[H 1];
Ytemp=H*W2';
%Using sigmoidal function
Y=logsig(Ytemp);
YCOL=[YCOL;Y(1) Y(2)];
%Updating weights W2
W2(1,:)=W2(1,:)+0.01*(TARGET(1,n)-Y(1)).*H;
W2(2,:)=W2(2,:)+0.01*(TARGET(2,n)-Y(2)).*H;
%Updating weights W1
G1=(TARGET(1,n)-Y(1)).*W2(1,:).*H.*(1-H);
G1=G1(1:1:1000);
G2=(TARGET(2,n)-Y(2)).*W2(2,:).*H.*(1-H);
G2=G2(1:1:1000);
W1=W1+0.01*TRAINDATA(n,:)'*(G1+G2);
```

```
end
JCOL=[JCOL -1*sum(sum(TARGET.*log(YCOL')))-1*
sum(sum((1-TARGET).*log((1-YCOL)')))];
end
YCOLTRAIN=YCOL;
figure
plot(JCOL)
title('Convergence of the Multi-class BPNN')
YCOLTEST=[];
for n=1:1:10
H=logsig(TESTDATA(n,:)*W1);
H=[H 1];
Ytemp=H*W2';
%Using sigmoidal function
Y=logsig(Ytemp);
YCOLTEST=[YCOLTEST;Y(1) Y(2)];
end
figure
subplot(2,1,1)
stem(YCOLTRAIN(:,1),'k-')
hold on
stem(YCOLTRAIN(:,2),'ro')
hold on
stem(TARGET(1,:),'k*')
hold on
stem(TARGET(2,:),'rs')

subplot(2,1,2)
stem(YCOLTEST(:,1),'k-')
hold on
stem(YCOLTEST(:,2),'ro')
hold on
stem(TARGET(1,:),'k*')
hold on
stem(TARGET(2,:),'rs')
```

5.6.1.2 ANN for Regression

In this case, purelinear is used as the activation function in the output layer, and the objective function is the sum-squared error as follows:

$$J = \sum_{n=1}^{n=N} \sum_{i=1}^{i=r} (y_{ni} - t_{ni})^2 \tag{5.23}$$

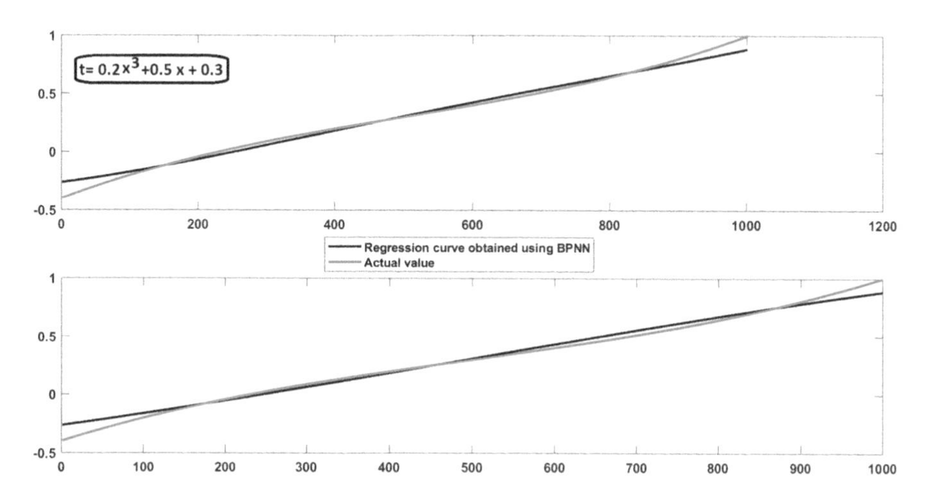

Fig. 5.18 Polynomial curve fitting obtained using BPNN-based neural network

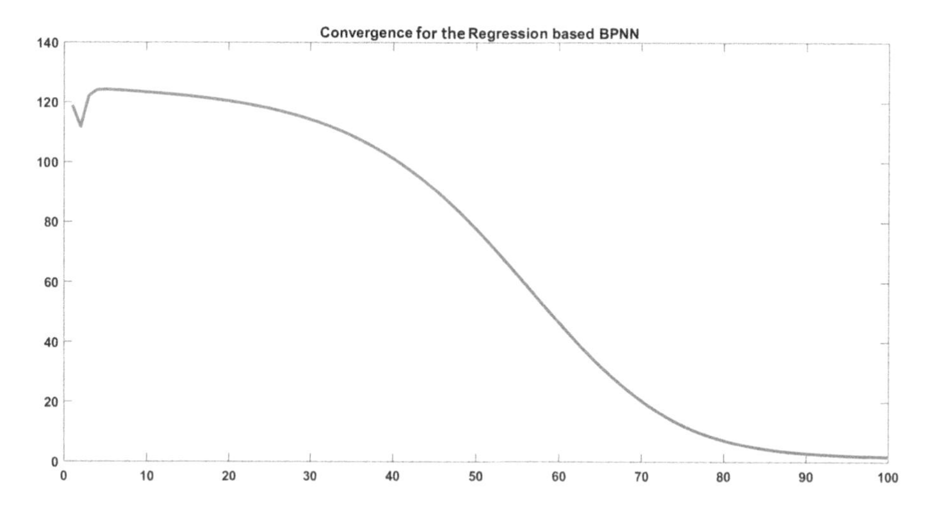

Fig. 5.19 Convergence graph achieved for the regression problem

In this case, gradient adjustment (weight update) follows (5.17)–(5.19) with y_i computed using (5.23). The polynomial curve fitting and corresponding convergence graph for the regression problem are illustrated in Figs. 5.18 and 5.19.

ANNregression.m

```
%ANNdemo for regression problem
x=-1:0.001:1;
t=0.2*(x.^3)+0.5*x+0.3;
TRAININPUT=x(1:2:length(x))';
TRAINOUTPUT=t(1:2:length(x))';
TESTINPUT=x(2:2:length(x))';
```

```
TESTOUTPUT=t(2:2:length(x))';
W1=randn(1,3)*0.1;
W2=randn(1,4)*0.1;
JCOL=[];
for iteration=1:1:100
   YCOLTRAIN=[];
for n=1:1:length(TRAININPUT)
H=logsig(TRAININPUT(n)*W1);
H=[H 1];
Y=H*W2';
YCOLTRAIN=[YCOLTRAIN;Y(1)];
%Updating weights W2
W2(1,:)=W2(1,:)+0.001*((TRAINOUTPUT(n)-Y(1))).*H;
%Updating weights W1
G1=(TRAINOUTPUT(n)-Y(1)).*W2(1,:).*H.*(1-H);
G1=G1(1:1:3);
W1=W1+0.001*TRAININPUT(n,:)'*G1;
end
JCOL=[JCOL sum((YCOLTRAIN-TRAINOUTPUT).^2)];
end
figure
plot(JCOL)
title('Convergence of the Regression based BPNN')
YCOLTEST=[];
for n=1:1:length(TESTINPUT)
H=logsig(TESTINPUT(n)*W1);
H=[H 1];
Y=H*W2';
%Using sigmoidal function
YCOLTEST=[YCOLTEST;Y];
end

figure
subplot(2,1,1)
plot(YCOLTRAIN,'k-')
hold on
plot(TRAINOUTPUT,'r-')
subplot(2,1,2)
plot(YCOLTEST,'k-')
hold on
plot(TESTOUTPUT,'r-')
```

5.6.1.3 ANN for Gaussian Mixture Model Estimation

In the case of regression, the observation is assumed as noisy observation of the form $t_n = y(x_n) + \epsilon$. In this, $y(x_n, w)$ is the parametric linear regression model relating x_n and t_n. It is usually assumed that the additive noise described by the random variable ϵ is multivariate Gaussian distributed with the mean zero and the co-variance matrix $\frac{1}{\beta}I$. If the noise is multimodel, the pdf is approximated using Gaussian Mixture Model (GMM). The unknown parameter describing the GMM is estimated using ANN as follows:

1. Let us assume the noise is having two modes and is modeled using mixtures of two Gaussian density function as follows:

$$f(x) = f_1(x)\alpha_1 + f_2(x)\alpha_2 \tag{5.24}$$

   where $f_1(x)$ is the Gaussian density function with mean μ_1 and variance σ_1 and $f_2(x)$ is the Gaussian density function with mean μ_2 and variance σ_2.
2. The Neural Network is modeled with six neurons in the output layer describing $o_1 = \mu_1, o_2 = \mu_2, \sigma_1 = e^{o_3}$, and $\sigma_2 = e^{o_4}$ and $\alpha_1 = \frac{o_1}{o_1+o_2}$ and $\alpha_2 = \frac{o_2}{o_1+o_2}$. The network is trained such that the following likelihood function is maximized.

$$J = f(x_1)f(x_2)\cdots f(x_N) \tag{5.25}$$

$$J = -\sum_{n=1}^{n=N} log(f(x_n)) \tag{5.26}$$

3. The number of neurons in the input layer is 1 (size of x is 1×1).
4. The training algorithm to update the variables associated with output neurons is summarized as follows:

   - Initialize the weight matrices.
   - For the input x_1, the corresponding output o_i for $i = 1 \cdots 6$ is obtained.
   - Compute $f_1(t_1)$ and $f_2(t_1)$, where t_1 is the corresponding target for the input x_1. Note that the function f_1 is described by the parameters α_1, μ_1, and σ_1. Also the function f_2 is described by the parameters α_2, μ_2, and σ_2.
   - Compute $\gamma_1 = \frac{\alpha_1 f_1(t_1)}{\alpha_1 f_1(t_1)+\alpha_2 f_2(t_1)}$ and $\gamma_2 = \frac{\alpha_2 f_2(t_1)}{\alpha_1 f_1(t_1)+\alpha_2 f_2(t_1)}$.
   - Compute the gradient in the output layer as follows:

$$\nabla o_1 = \alpha_1 - \gamma_1 \tag{5.27}$$

$$\nabla o_2 = \alpha_2 - \gamma_2 \tag{5.28}$$

$$\nabla o_3 = \gamma_1(\frac{o_1 - t_1}{\sigma_1^2}) \tag{5.29}$$

$$\nabla o_4 = \gamma_2(\frac{o_2 - t_2}{\sigma_2^2}) \tag{5.30}$$

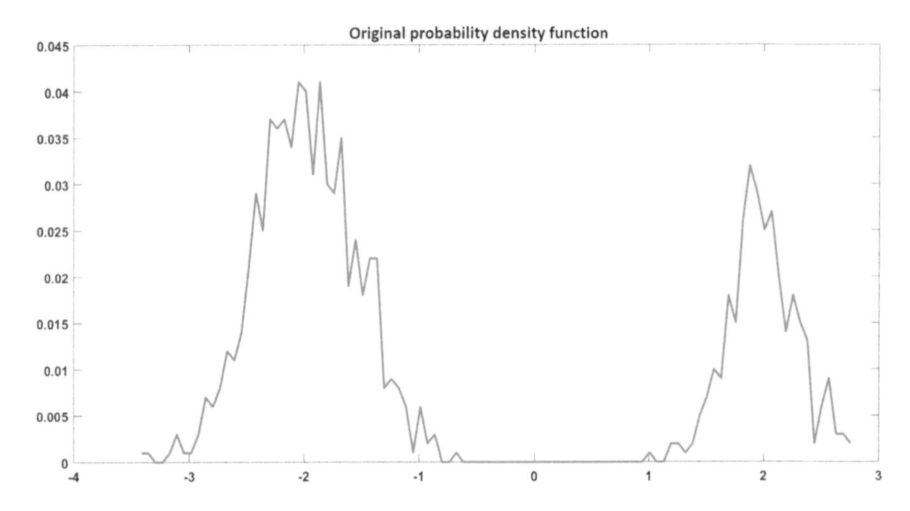

Fig. 5.20 Original GMM pdf

$$\nabla o_5 = -\gamma_1(\frac{(t_1 - o_1)^2}{\sigma_1^2} - \frac{1}{\sigma_1}) \tag{5.31}$$

$$\nabla o_6 = -\gamma_2(\frac{(t_2 - o_2)^2}{\sigma_2^2} - \frac{1}{\sigma_2}) \tag{5.32}$$

- These are treated as $e_1, e_2, ..., e_6$, and (5.17)–(5.19) are used to update the weight matrix.

The illustration on estimation of GMM convergence graph is illustrated in Figs. 5.20 and 5.21. Also the pdf of original GMM is given in Fig. 5.22 for comparison.

BPNN-GMM.m

```
%Estimating GMM parameters using BPNN
m1=2;
m2=-2;
v1=0.1;
v2=0.2;
p1=0.3;
p2=0.7;
ORIGINAL=[m1 m2 v1 v2 p1 p2];
x=ones(1,1000);
for i=1:1:length(x)
r=rand;
    if(r>p1)
        t(i)=randn*sqrt(v2)+m2;
```

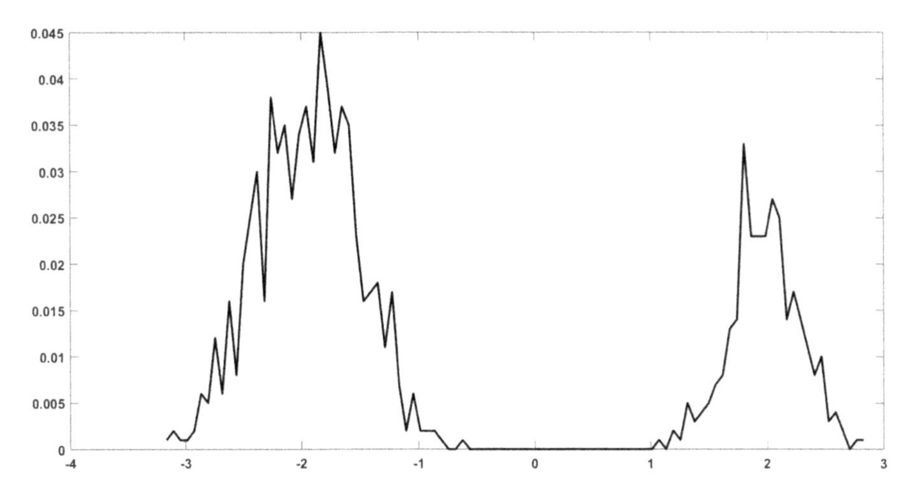

Fig. 5.21 Estimated GMM pdf

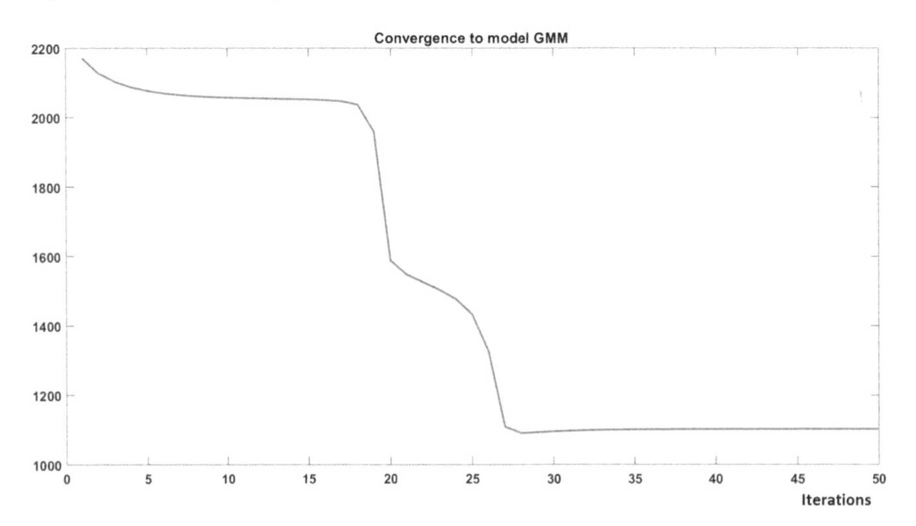

Fig. 5.22 Convergence graph achieved for GMM estimation

```
        else
            t(i)=randn*sqrt(v1)+m1;
        end
    end
    [t1,x1]=hist(t,100);
    t1=t1/sum(t1);
    figure(1)
    plot(x1,t1,'r')
    %Number of neurons in the output layer is given
```

```
%as 6 and the corresponding activation functions are
given below
%[m1 m2 exp(v1) exp(v2) exp(p1)/(exp(p1)+exp(p2))
%exp(p2)/(exp(p1)+exp(p2))] the following
%%%%%%%%%%%%%%%%%%%%%%%%%%%%%%%%%%%%%%%%%%%%%% %%
TRAININPUT=x';
T  RAINOUTPUT=t';
W1=randn(1,100)*0.0001;
W2=randn(6,101)*0.0001;
JCOL=[];
for iteration=1:1:50
   YCOLTRAIN=[];
for n=1:1:length(TRAININPUT)
H=logsig(TRAININPUT(n)*W1);
H=[H 1];
Y=H*W2';
%Applying activation
Y(5)=(exp(Y(5)))/(exp(Y(5))+exp(Y(6)));
Y(6)=(exp(Y(6)))/(exp(Y(5))+exp(Y(6)));
Y(3)=exp(Y(3));
Y(4)=exp(Y(4));
%Updating weights W2
temp1=normpdf(TRAINOUTPUT(n),Y(1),Y(3));
temp2=normpdf(TRAINOUTPUT(n),Y(2),Y(4));
gamma1=Y(5)*temp1/(Y(5)*temp1+Y(6)*temp2);
gamma2=Y(6)*temp2/(Y(5)*temp1+Y(6)*temp2);
GRADIENT=[gamma1*((Y(1)-TRAINOUTPUT(n))/(Y(3)^2))...
          gamma2*((Y(2)-TRAINOUTPUT(n))/(Y(4)^2)) ...
          -gamma1*(((TRAINOUTPUT(n)-Y(1))^2)/(Y(3)^3)
          -(1/Y(3)))...
          -gamma2*(((TRAINOUTPUT(n)-Y(2))^2)/(Y(4)^3)
          -(1/Y(4)))...
          (Y(5)-gamma1) (Y(6)-gamma2)];
W2(1,:)=W2(1,:)-0.0001*GRADIENT(1).*H;
W2(2,:)=W2(2,:)-0.0001*GRADIENT(2).*H;
W2(3,:)=W2(3,:)-0.0001*GRADIENT(3).*H;
W2(4,:)=W2(4,:)-0.0001*GRADIENT(4).*H;
W2(5,:)=W2(5,:)-0.0001*GRADIENT(5).*H;
W2(6,:)=W2(6,:)-0.0001*GRADIENT(6).*H;
%Updating weights W1
G1=GRADIENT(1).*W2(1,:).*H.*(1-H);
G1=G1(1:1:100);
G2=GRADIENT(2).*W2(2,:).*H.*(1-H);
G2=G2(1:1:100);
G3=GRADIENT(3).*W2(3,:).*H.*(1-H);
```

```
G3=G3(1:1:100);
G4=GRADIENT(4).*W2(4,:).*H.*(1-H);
G4=G4(1:1:100);
G5=GRADIENT(5).*W2(5,:).*H.*(1-H);
G5=G5(1:1:100);
G6=GRADIENT(6).*W2(6,:).*H.*(1-H);
G6=G6(1:1:100);
W1=W1-0.0001*TRAININPUT(n,:)'*(G1+G2+G3+G4+G5+G6);
end
s=0;
for n=1:1:length(TRAININPUT)
temp1=normpdf(TRAINOUTPUT(n),Y(1),Y(3));
temp2=normpdf(TRAINOUTPUT(n),Y(2),Y(4));
s=s-log(Y(5)*temp1+Y(6)*temp2);
end
JCOL=[JCOL s];
end
figure(2)
plot(JCOL)
title('Convergence of the Regression based BPNN')
%%%%%%%%%%%%%%%%%%%%%%%%%%%%%%%%%%%
H=logsig(W1);
H=[H 1];
Y=H*W2';
Y(5)=exp(Y(5))/(exp(Y(5))+exp(Y(6)));
Y(6)=exp(Y(6))/(exp(Y(5))+exp(Y(6)));
Y(3)=exp(Y(3));
Y(4)=exp(Y(4));
ORIGINAL=[m1 m2 v1 v2 p1 p2];
ESTIMATED=[Y(1) Y(2) Y(3)^2 Y(4)^2 Y(5) Y(6)];
[ORIGINAL;ESTIMATED]
figure
DATA=[];
 for i=1:1:1000
    r=rand;
    if(r>Y(5))
        DATA=[DATA randn*Y(4)+Y(2)];
    else
        DATA=[DATA randn*Y(3)+Y(1)];
    end
 end
[t,x]=hist(DATA,100);
t=t/sum(t);
figure(3)
plot(x,t,'k')
```

5.6.2 Kernel Trick–Based Class-Conditional Probability Density Function Using Neural Network

In this technique, for an arbitrary vector, **x**, the class conditional probability density function $p(\mathbf{x}/c_k)$ is empirically computed (refer (5.33)) as the ratio of the sum of Euclidean distances between the vector **x** and the centroid of the kth class. BPNN is trained with **x** as the input and the corresponding $p(\mathbf{x}/c_k)$ as the target value. Individual network is trained to obtain class conditional pdf of the individual classes. The test vector is given as the input to the individual ANN, and the index of the ANN that has the maximum value is declared as the class index. Two-class problem using kernel trick–based ccpdf using Neural Network is illustrated in Figs. 5.23, 5.24, 5.25, and 5.26.

$$p(\mathbf{x}/c_k) = K \sum_{i=1}^{i=r,i\neq k} \frac{d_{\mathbf{x},c_i}}{d_{\mathbf{x},c_k}} \tag{5.33}$$

where K is the constant that satisfies the condition $\sum_{k=1}^{k=K} p(c_k/\mathbf{x}) = 1$. The Euclidean distances are computed in the higher dimensional space. The distance between the ith vector in the jth class with the centroid of the qth class is given as follows:

$$k(u_{ij}, u_{ij}) - \left(\frac{2}{n_q}\right) \sum_{k=1}^{n_q} k(u_{ij}, u_{kq}) + \frac{1}{n_q^2} \sum_{k=1}^{k=n_q} \sum_{l=1}^{l=n_q} k(u_{kq}, u_{lq}) \tag{5.34}$$

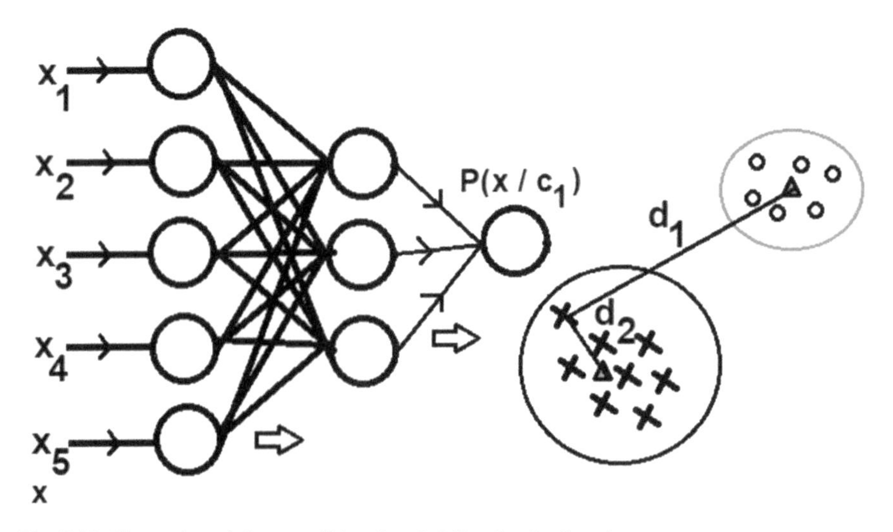

Fig. 5.23 Illustration of class conditional probability density function

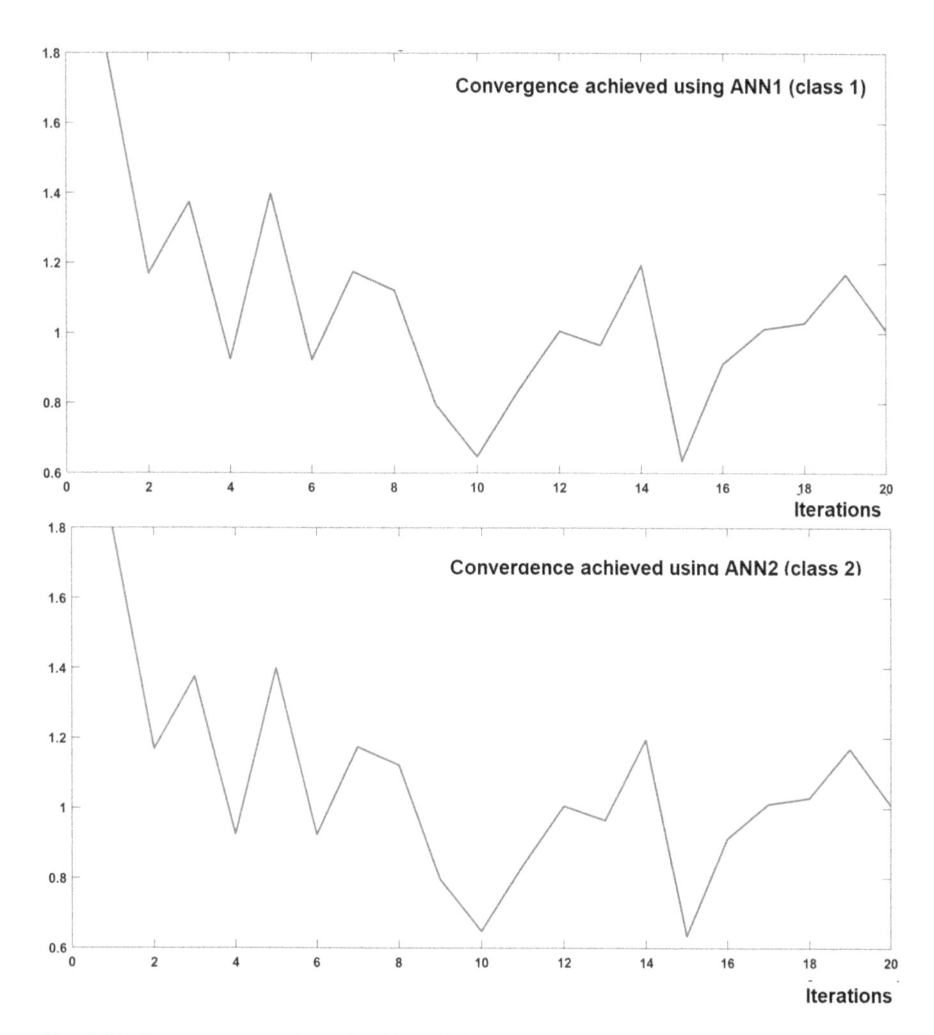

Fig. 5.24 Convergence graph obtained by training ANN1 (class 1) and ANN2 (class 2)

ccpdf.m

```
clear all
close all
%ccpdf
load IMAGEDATA
%The euclidean distances of the individual vectors of
class 1
%with the centroid of the class 1 and class 2 in the
higher
%dimensional space is computed as follows.
%Number of vectors in each classes are given as 10.
```

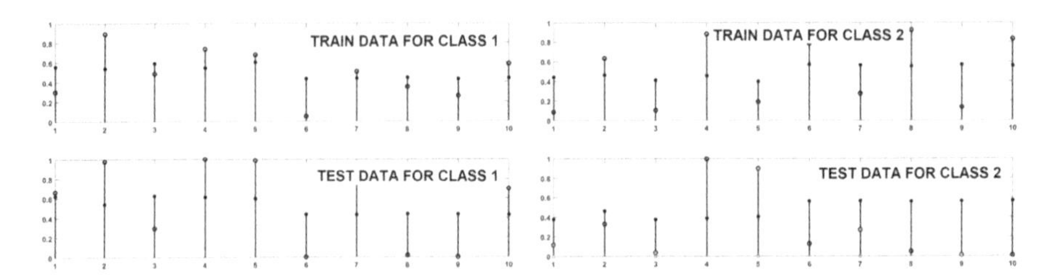

Fig. 5.25 Class index probability for the training and testing data for class 1 and class 2 (ANN 1 and ANN2) and the corresponding attained values using trained ANN 1 and ANN 2

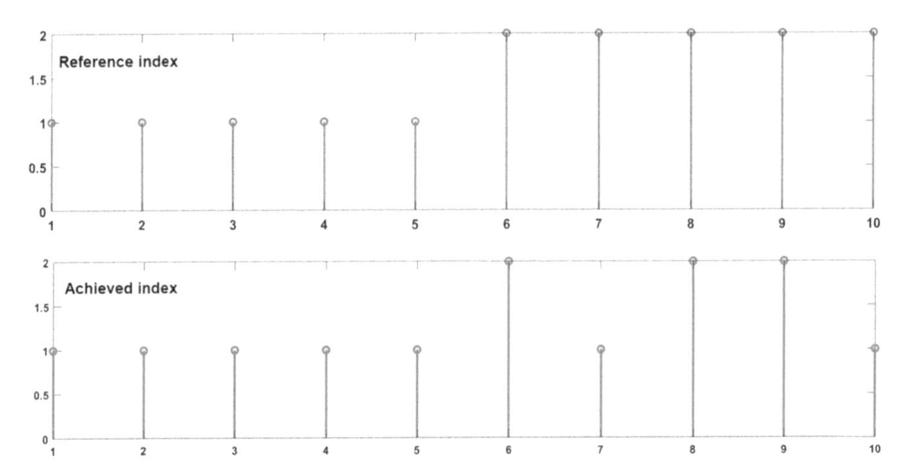

Fig. 5.26 Original class label and the class label achieved using the trained networks (ANN1 and ANN2). It is seen that 90% success rate is achieved using the constructed model

```
sigma=5;
for i=1:1:10
   for j=1:1:10
      temp{1}(i,j)=gbf(CLASS1{i},CLASS1{j},sigma);
   end
end

for i=1:1:10
   for j=1:1:10
      temp{2}(i,j)=gbf(CLASS2{i},CLASS2{j},sigma);
   end
end
for i=1:1:10
    for j=1:1:10
         temp3(i,j)=gbf(CLASS1{i},CLASS2{j},sigma);
    end
end
```

```
%Distance between the vectors in the first class
%with centroid of the first class
  for i=1:1:10
      for j=1:1:2
          for q=1:1:2
              if(j~=q)
              d(i,j,q)=temp{j}(i,i)-(2/10)*
              sum(temp3(q,:))...
              +(1/100)* sum(sum(temp{q}))
              else
              d(i,j,q)=temp{j}(i,i)-(2/10)*
              sum(temp{j}(i,:))...
              +(1/100)* sum(sum(temp{q}))
              end
          end
       end
  end
 %The posterior density function is computed
 %as following
 NUM=0
 for i=1:1:10
     for j=1:1:2
         for q=1:1:2
             NUM=NUM+d(i,j,q);
         end
         t1=NUM/d(i,j,1);
         t2=NUM/d(i,j,2 );
         PX1(i,j)=t1/(t1+t2);
         PX2(i,j)=t2/(t1+t2);
     end
 end
%ANN 1 is trained using elements of the matrix PX1
%as the target values corresponding to the input
%CLASS{1} and CLASS{2} to model posterior density
function of
%class 1
temp1=cell2mat(CLASS1');
temp2=cell2mat(CLASS2');
TRAINDATA=[temp1(1:1:5,:) ones(5,1);temp2(1:1:5,:)
ones(5,1)];
TESTDATA=[temp1(6:1:10,:) ones(5,1);temp2(6:1:10,:)
ones(5,1)];
temp=reshape(PX1,1,20);
TRAINTARGET=[temp([1:1:5]) temp([11:1:15])];
TESTTARGET=[temp([6:1:10]) temp([16:1:20])];
```

```
W1=randn(2501,1000)*0.5;
W2=randn(1,1001)*0.5;
JCOL=[];
for iteration=1:1:20
   YCOLTRAIN=[];
    for n=1:1:10
H=logsig(TRAINDATA(n,:)*W1);
H=[H 1];
Y=H*W2';
Y=1/(1+exp(-Y));
YCOLTRAIN=[YCOLTRAIN;Y(1)];
%Updating weights W2
W2(1,:)=W2(1,:)+0.01*(TRAINTARGET(1,n)-Y(1)).*H;
%Updating weights W1
G1=(TRAINTARGET(1,n)-Y(1)).*W2(1,:).*H.*(1-H);
G1=G1(1:1:1000);
W1=W1+0.01*TRAINDATA(n,:)'*(G1);
end
JCOL=[JCOL sum((TRAINTARGET-YCOLTRAIN').^2)];
end
figure
plot(JCOL)
title('Convergence of the Multi-class BPNN')
YCOLTEST=[];
for n=1:1:10
H=logsig(TESTDATA(n,:)*W1);
H=[H 1];
Y=H*W2';
Y=1/(1+exp(-Y));
YCOLTEST=[YCOLTEST;Y];
end
figure
subplot(2,1,1)
stem(YCOLTRAIN,'k-')
hold on
stem(TRAINTARGET,'k*')
subplot(2,1,2)
stem(YCOLTEST,'k-')
hold on
stem(TESTTARGET,'k*')
PART1=YCOLTEST;
%In the same fashion, ANN 2 is trained using elements
of the matrix PX2
%as the target values corresponding to the input
%CLASS{1} and CLASS{2} to model posterior density
```

```
function of
%class 2
temp1=cell2mat(CLASS1');
temp2=cell2mat(CLASS2');
TRAINDATA=[temp1(1:1:5,:) ones(5,1);temp2(1:1:5,:)
ones(5,1)];
TESTDATA=[temp1(6:1:10,:) ones(5,1);temp2(6:1:10,:)
ones(5,1)];
temp=reshape(PX2,1,20);
TRAINTARGET=[temp([1:1:5]) temp([11:1:15])];
TESTTARGET=[temp([6:1:10]) temp([16:1:20])];
W1=randn(2501,1000)*0.5;
W2=randn(1,1001)*0.5;
JCOL=[];
for iteration=1:1:20
   YCOLTRAIN=[];
    for n=1:1:10
H=logsig(TRAINDATA(n,:)*W1);
H=[H 1];
Y=H*W2';
Y=1/(1+exp(-Y));
YCOLTRAIN=[YCOLTRAIN;Y(1)];
%Updating weights W2
W2(1,:)=W2(1,:)+0.01*(TRAINTARGET(1,n)-Y(1)).*H;
%Updating weights W1
G1=(TRAINTARGET(1,n)-Y(1)).*W2(1,:).*H.*(1-H);
G1=G1(1:1:1000);
W1=W1+0.01*TRAINDATA(n,:)'*(G1);
end
JCOL=[JCOL sum((TRAINTARGET-YCOLTRAIN').^2)];
end
figure
plot(JCOL)
title('Convergence of the Multi-class BPNN')
YCOLTEST=[];
for n=1:1:10
H=logsig(TESTDATA(n,:)*W1);
H=[H 1];
Y=H*W2';
Y=1/(1+exp(-Y));
YCOLTEST=[YCOLTEST;Y];
end

figure
subplot(2,1,1)
```

```
stem(YCOLTRAIN,'k-')
hold on
stem(TRAINTARGET,'k*')
subplot(2,1,2)
stem(YCOLTEST,'k-')
hold on
stem(TESTTARGET,'k*')
PART2=YCOLTEST;
[p,q]=max([PART1 PART2]')
REF=[ones(1,5) ones(1,5)*2];
POS=length(find((REF-q)==0))*10;
figure
subplot(2,1,1)
stem(REF)
subplot(2,1,2)
stem(q,'r')
```

5.6.3 Convolution Neural Network (CNN)

Convolution neural network is the feed-forward network with shared weights (Fig. 5.27). For the 4×4 input pattern, 3×3 filter is used as the shared weights. This helps to collect the features (4×4) from the pattern. This is further fully connected to obtain the intended target. Sigmoidal transfer function is used in the

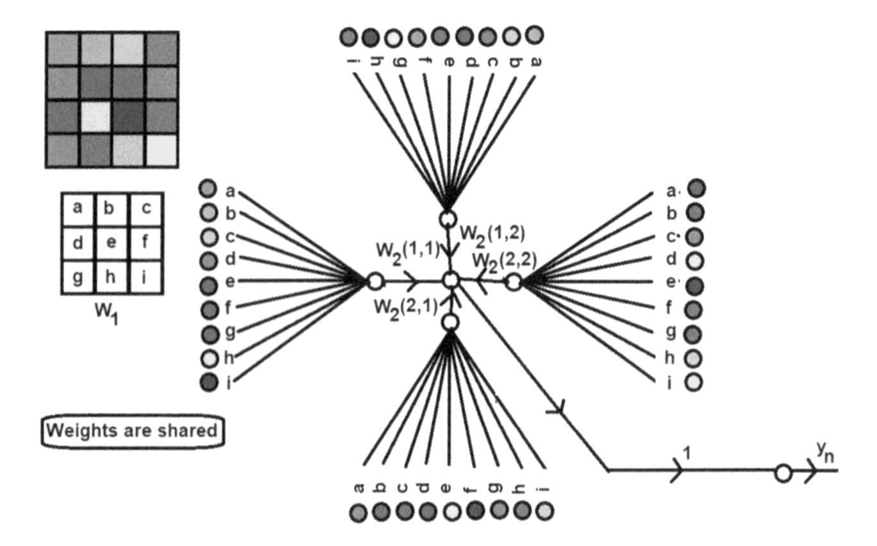

Fig. 5.27 Illustration of CNN with *a*, *b*, *c*, ..., *i* as the shared weights to classify 4×4 pattern into one among the two categories

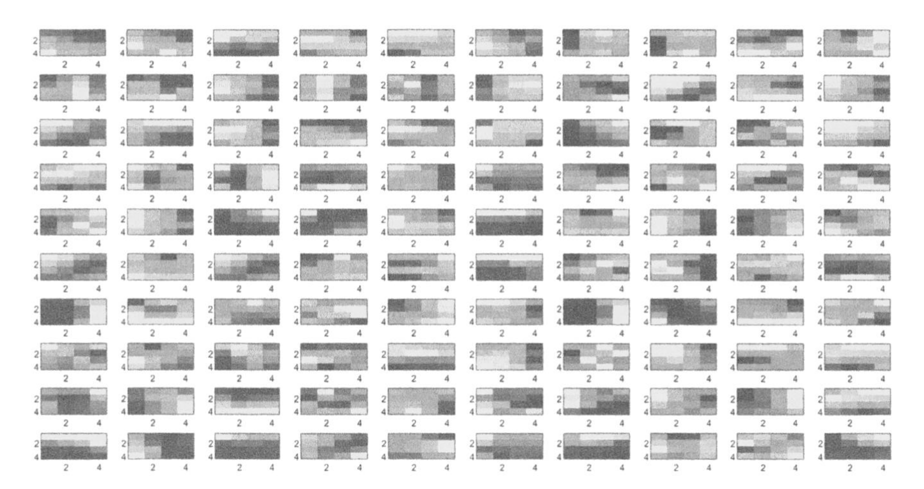

Fig. 5.28 Training set for class 1

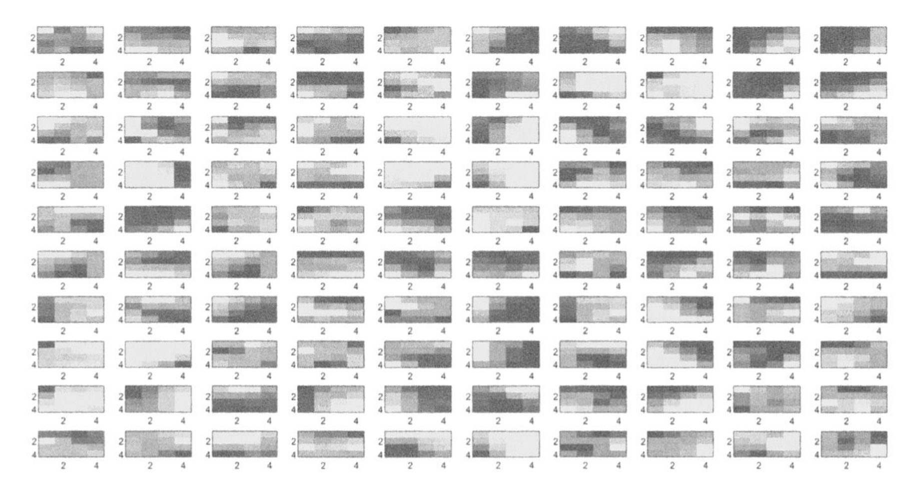

Fig. 5.29 Training set for class 2

hidden and the output layer. The cross-entropy objective function is used and the gradient technique is used to update the filter coefficients (shared weights). The convergence achieved using the model is given in Fig. 5.28. The training and the testing data for class 1 and class 2 are given in Figs. 5.28, 5.29, 5.30, and 5.31. The intended target values along with the attained target values are also illustrated in Fig. 5.32. In general, more than one layer can be used. Also in each layer, more than one filter can be used. For the multi-class problem, soft-max function is used in the output layer. Downsampling is done at every layer (to reduce the computation time and complexity), and pooling techniques are used to avoid over fitting. The corresponding already trained network is connected to fully connected network to perform the intended multi-class problem. This is known as transfer learning. For

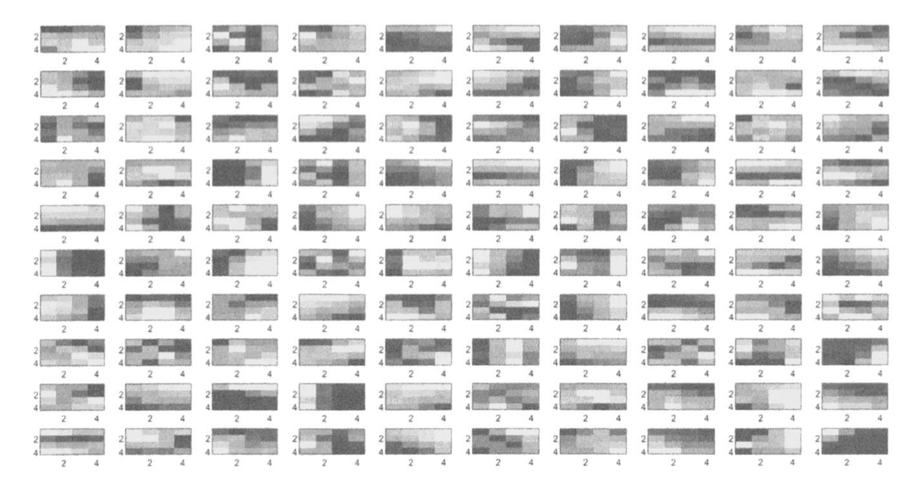

Fig. 5.30 Testing set for class 1

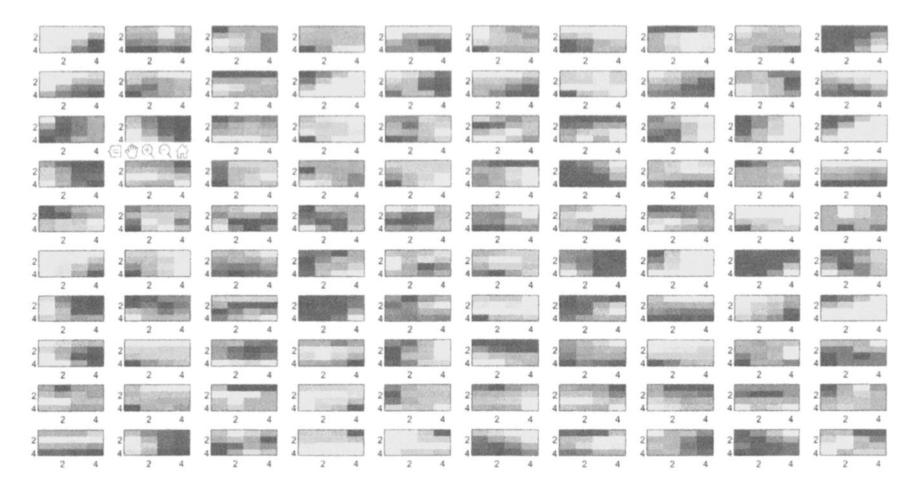

Fig. 5.31 Testing set for class 2

instance, trained CNN for the vehicle detection may be used to connect with the fully connected network to construct the classifier to detect subcategory like Road vehicle versus Train. The proper combination of the CNN with Recurrent Neural Network (RNN), Generative Adversarial Network, LSTM (Long–Short Term Memory), etc. can be used to construct new model for the specified applications.

convolution-network.m

```
load DATACONV
figure
for i=1:1:100
figure(1)
```

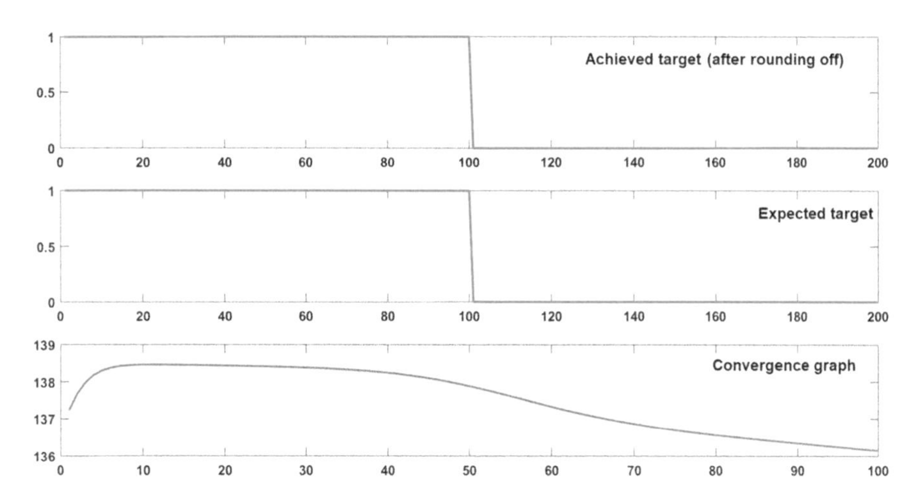

Fig. 5.32 Illustration of the convolution network

```
subplot(10,10,i)
imagesc(TRAINDATA1{i})
figure(2)
subplot(10,10,i)
imagesc(TRAINDATA2{i})
figure(3)
subplot(10,10,i)
imagesc(TESTDATA1{i})
figure(4)
subplot(10,10,i)
imagesc(TESTDATA2{i})
end

TRAINTARGET=[ones(1,100) zeros(1,100)];
TESTTARGET=[ones(1,100) zeros(1,100)];
gamma=0.01;
%Initializing weights
W1=randn(3,3)*0.1;
W2=randn(2,2)*0.1;
J=[];
for epoch=1:1:100
OUT=[];
for iteration=1:1:100
fun = @(x) 1/(1+exp(-sum(sum(x.*W1))));
A=blkproc(TRAINDATA1{iteration},[3 3],fun);
out=1/(1+exp(-sum(sum(A.*W2))));
OUT=[OUT out];
```

```
W2=W2+gamma*(TRAINTARGET(iteration)-out)*A;
%Update the weights W1 and W2 as follows.
temp=A.*(A-1);
%fun = @(x) cell2mat(x);
%TEMP=blkproc(TRAINDATA1{iteration},[3 3],fun);
TEMP=TRAINDATA1{iteration};
W1=W1+gamma*temp(1,1)*TEMP(1:1:3,1:1:3)*W2(1,1)*...
(TRAINTARGET(iteration)-out);
W1=W1+gamma*temp(1,2)*TEMP(1:1:3,2:3:4)*W2(1,2)*...
(TRAINTARGET(iteration)-out);
W1=W1+gamma*temp(2,1)*TEMP(2:1:4,1:1:3)*W2(2,1)*...
(TRAINTARGET(iteration)-out);
W1=W1+gamma*temp(2,2)*TEMP(2:1:4,2:1:4)*W2(2,2)*...
(TRAINTARGET(iteration)-out);
end
for iteration=1:1:100
fun = @(x) 1/(1+exp(-sum(sum(x.*W1))));
A=blkproc(TRAINDATA2{iteration},[3 3],fun);
out=1/(1+exp(-sum(sum(A.*W2))));
OUT=[OUT out];
W2=W2+gamma*(TRAINTARGET(iteration+100)-out)*A;
%Update the weights W1 and W2 as follows.
temp=A.*(1-A);
%fun = @(x) cell2mat(x);
%TEMP=blkproc(TRAINDATA1{iteration},[3 3],fun);
TEMP=TRAINDATA2{iteration};
W1=W1+gamma*temp(1,1)*TEMP(1:1:3,1:1:3)*W2(1,1)*...
(TRAINTARGET(iteration+100)-out);
W1=W1+gamma*temp(1,2)*TEMP(1:1:3,2:3:4)*W2(1,2)*...
(TRAINTARGET(iteration+100)-out);
W1=W1+gamma*temp(2,1)*TEMP(2:1:4,1:1:3)*W2(2,1)*...
(TRAINTARGET(iteration+100)-out);
W1=W1+gamma*temp(2,2)*TEMP(2:1:4,2:1:4)*W2(2,2)*...
(TRAINTARGET(iteration+100)-out);
end
s1=sum(TRAINTARGET(1:1:100).*log(OUT(1:1:100)));
s2=sum((1-TRAINTARGET(101:1:200)).*log(1-OUT
(101:1:200)));
J=[J -(s1+s2)];
end

%Testing data
OUTTEST=[]
for i=1:1:100
fun = @(x) 1/(1+exp(-sum(sum(x.*W1))));
```

```
A=blkproc(TESTDATA1{iteration},[3 3],fun);
out=1/(1+exp(-sum(sum(A.*W2))));
OUTTEST=[OUTTEST out];
end
for i=1:1:100
fun = @(x) 1/(1+exp(-sum(sum(x.*W1))));
A=blkproc(TESTDATA2{iteration},[3 3],fun);
out=1/(1+exp(-sum(sum(A.*W2))));
OUTTEST=[OUTTEST out];
end
figure
subplot(3,1,1)
plot(round(OUTTEST),'r')
subplot(3,1,2)
plot(TESTTARGET,'r')
subplot(3,1,3)
plot(J)
```

5.6.4 Generative Adversarial Network

In this technique, two blocks, namely generator block and the discriminator block, are used. The input to the generator block is the outcome of the random variable z, which follows probability density function p_z. The output of the discriminator is the outcome of the random variable x_{gen} that follows the pdf p_g. The weights of the generator block are trained, such that after training, the outcome x_{gen} follows the pdf of x_{ori} represented as p_{ori}. The outcome of the discriminator is the sigmoid function that assigns 1 if the input to the discriminator block is the outcome of the random variable x_{ori}. Similarly, the target is assigned as 0, if the input to the discriminator is the outcome of the random variable not belonging to x_{ori}. This is known as adversarial network. The objective function used to train GAN network is given as follows:

$$J_1 = -\sum_{n=1}^{n=N} t_n ln(y_n) \tag{5.35}$$

$$J_2 = -\sum_{n=1}^{n=N} (1 - t_n) ln((1 - t_n)) \tag{5.36}$$

The discriminator network is trained by minimizing $J_1 + J_2$, and the generator network is trained by maximizing J_2. The training equations are computed. Let the weights connecting the input and the hidden layer (j^{th} the neuron of the input layer with k^{th} neuron of the hidden layer) of the generator network are represented as w^{α}_{jk}. Similarly, the weights connecting the hidden layer (j^{th} the neuron of the hidden layer with k^{th} neuron of the output layer) are represented as v^{α}_{jk}. Similarly, the corresponding weights used in the discriminator network are represented as w^{β}_{jk} and v^{β}_{jk}, respectively. The updated equations are given below. The GAN algorithm is illustrated in Figs. 5.33 and 5.34.

$$\frac{\partial(J_1 + J_2)}{\partial v_{\beta}^{11}} = (t - y)h_{\beta}^{1} \tag{5.37}$$

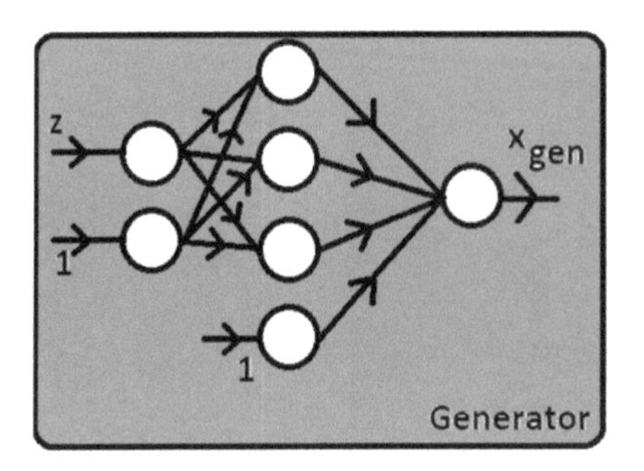

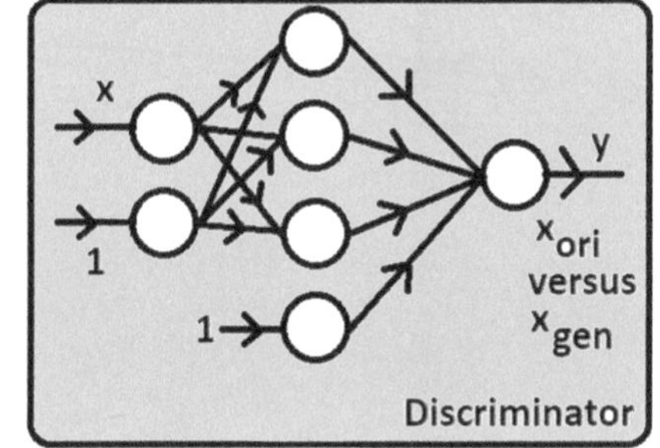

Fig. 5.33 Illustration of Generative Adversarial Network

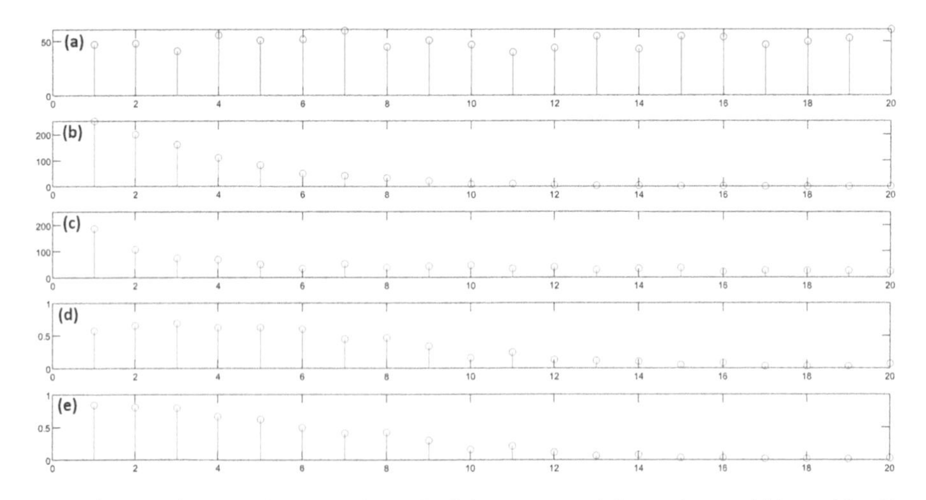

Fig. 5.34 (**a**) Histogram (p_z approximated) of the outcome of the random variable z with pdf p_z (input data). (**b**) Histogram of the outcome of the intended random variable x with p_{data} (approximated). (**c**) Histogram of the generated data with the intended random variable p_{data} generated using the Generator. (d) $\frac{p_{data}}{p_{data}+p_{gen}}$. (e) $\frac{p_{data}}{p_z}$. It is noted that p_z is uniformly distributed. p_{gen} follows p_{data}, i.e., exponentially distributed. First few bins of the subplot (d) reach the value 0.5 when compared with the value in subplot (e)

$$\frac{\partial(J_1 + J_2)}{\partial v_\beta^{21}} = (t - y)h_\beta^2 \tag{5.38}$$

$$\frac{\partial(J_1 + J_2)}{\partial v_\beta^{31}} = (t - y)h_\beta^3 \tag{5.39}$$

$$\frac{\partial(J_1 + J_2)}{\partial v_\beta^{41}} = (t - y)h_\beta \tag{5.40}$$

$$\frac{\partial(J_1 + J_2)}{\partial w_\beta^{11}} = (t - y)(h_\beta^1)(1 - h_\beta^1)x \tag{5.41}$$

$$\frac{\partial(J_1 + J_2)}{\partial w_\beta^{12}} = (t - y)(h_\beta^1)(1 - h_\beta^2)x \tag{5.42}$$

$$\frac{\partial(J_1 + J_2)}{\partial w_\beta^{13}} = (t - y)(h_\beta^1)(1 - h_\beta^3)x \tag{5.43}$$

$$\frac{\partial(J_1 + J_2)}{\partial w_\beta^{21}} = (t - y)(h_\beta^1)((t - y)1 - h_\beta^1) \tag{5.44}$$

$$\frac{\partial(J_1 + J_2)}{\partial w_\beta^{22}} = (t - y)(h_\beta^1)(1 - h_\beta^2) \tag{5.45}$$

$$\frac{\partial(J_1 + J_2)}{\partial w_\beta^{23}} = (t - y)(h_\beta^1)(1 - h_\beta^3) \tag{5.46}$$

$$temp = v_\beta^1 h_\beta^1 (1 - h_\beta^1) w_\beta^1 + v_\beta^2 (1 - h_\beta^2) h_\beta^2 w_\beta^2 + v_\beta^3 (1 - h_\beta^3) h_\beta^3 w_\beta^3 \tag{5.47}$$

$$\frac{\partial J_2}{\partial v_\alpha^{11}} = temp \times (1 - t) y x_{gen} (1 - x_{gen}) h_\alpha^1 \tag{5.48}$$

$$\frac{\partial J_2}{\partial v_\alpha^{21}} = temp \times (1 - t) y x_{gen} (1 - x_{gen}) h_\alpha^2 \tag{5.49}$$

$$\frac{\partial J_2}{\partial v_\alpha^{31}} = temp \times (1 - t) y x_{gen} (1 - x_{gen}) h_\alpha^3 \tag{5.50}$$

$$\frac{\partial J_2}{\partial v_\alpha^{41}} = temp \times (1 - t) y x_{gen} (1 - x_{gen}) \tag{5.51}$$

$$\frac{\partial J_2}{\partial w_\alpha^{11}} = temp \times (1 - t) y x_{gen} (1 - x_{gen}) h_\alpha^1 (1 - h_\alpha^1) z \tag{5.52}$$

$$\frac{\partial J_2}{\partial w_\alpha^{12}} = temp \times (1 - t) y x_{gen} (1 - x_{gen}) h_\alpha^2 (1 - h_\alpha^2) z \tag{5.53}$$

$$\frac{\partial J_2}{\partial w_\alpha^{13}} = temp \times (1 - t) y x_{gen} (1 - x_{gen}) h_\alpha^3 (1 - h_\alpha^3) z \tag{5.54}$$

$$\frac{\partial J_2}{\partial w_\alpha^{11}} = temp \times (1-t)yx_{gen}(1-x_{gen})h_\alpha^1(1-h_\alpha^1) \quad (5.55)$$

$$\frac{\partial J_2}{\partial w_\alpha^{12}} = temp \times (1-t)yx_{gen}(1-x_{gen})h_\alpha^2(1-h_\alpha^2) \quad (5.56)$$

$$\frac{\partial J_2}{\partial w_\alpha^{13}} = temp \times (1-t)yx_{gen}(1-x_{gen})h_\alpha^3(1-h_\alpha^3) \quad (5.57)$$

GANdemo.m

```
clear all
close all
%Generator
z=[rand(1,1000)*2-1;ones(1,1000)]';
Walpha1=randn(2,50)*0.5;
Halpha=logsig(z*Walpha1);
Halpha=[Halpha ones(1000,1)];
Walpha2=randn(51,1)*0.5;
xcap=logsig(Halpha*Walpha2);
xori=random('exp',1,1,1000);
xori=xori/max(xori);
x=[xori xcap'];
t=[ones(1,1000) zeros(1,1000)];
[p,q]=sort(rand(1,2000));
x=x(q);
t=t(q);
%Update weights based on the output
%from the discriminator as follows.
Wbeta1=randn(2,50)*1;
Wbeta2=randn(51,1)*1;
xmod=[x;ones(1,2000)];
for iteration=1:1:50
%Updating beta for discriminator
for iteration1=1:1:2
ycol=[];
for n=1:1:2000
Hbeta=logsig(Wbeta1'*xmod(:,n));
Hbeta=[Hbeta; 1];
y=logsig(Hbeta'*Wbeta2);
ycol=[ycol y];
Wbeta2=Wbeta2+0.01*[t(n)-y]*Hbeta;
G1=(t(n)-y)*Wbeta2(:,1).*Hbeta.*(1-Hbeta);
G1=G1(1:1:50);
Wbeta1=Wbeta1+0.01*xmod(:,n)*G1';
end
```

```
[mean(ycol(1:1:1000)) mean(ycol(1001:1:2000))]
end
%Updating alpha for generator
for iteration2=1:1:2
xcap=[];
for n=1:1:1000
Halpha=logsig(Walpha1'*z(n,:)');
Halpha=[Halpha; 1];
temp=logsig(Halpha'*Walpha2);
xcap=[xcap temp];
Hbeta=logsig(Wbeta1'*[temp 1]');
Hbeta=[Hbeta; 1];
y=logsig(Hbeta'*Wbeta2);
temp1=Hbeta(1:1:50);
temp2=y*(1-t(n+1000))*sum(temp1.*Wbeta2(1:50,1).*
Wbeta1(1,:)');
Walpha2=Walpha2+0.7*(Halpha'*temp2)'*temp*(1-temp);
G1=temp2*Walpha2(:,1).*Halpha.*(1-Halpha)*temp*
(1-temp);
G1=G1(1:1:50);
Walpha1=Walpha1+0.7*[z(n,:)]'*G1';
end
end
Halpha=logsig(z*Walpha1);
Halpha=[Halpha ones(1000,1)];
xcap=logsig(Halpha*Walpha2);
figure(1)
subplot(2,1,1)
plot(xori)
subplot(2,1,2)
plot(xcap,'r')
x=[xori xcap'];
xmod=[x;ones(1,2000)];
figure(2)
subplot(5,1,2)
[a1,b1]=hist(xori,20);
stem(a1)
xcap=xcap/max(xcap);
subplot(5,1,3)
[a2,b2]=hist(xcap,20);
stem(a2)
subplot(5,1,4)
stem(a1./(a1+a2))
subplot(5,1,1)
[a4,b4]=hist(z(:,1),20);
```

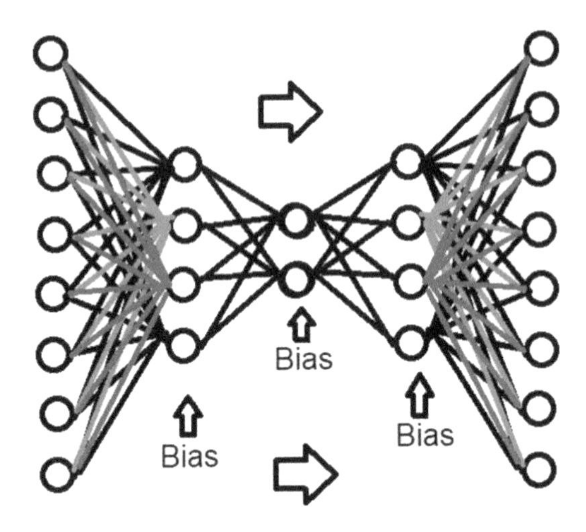

Fig. 5.35 Illustration of autoencoder network

```
pause(0.5)
stem(a4,'r')
subplot(5,1,5)
[a5,b5]=hist(z(:,1),20);
pause(0.5)
stem(a1./(a1+a5))
end
```

5.6.5 *Autoencoder Network*

It is the typical feed-forward network with target vectors that are identical as that of the input vectors. This network is used to represent the input vector with less number of elements (typically used for data compression). The illustration of autoencoder network is given in Figs. 5.35, 5.36, 5.37, and 5.38.

autoencoderdemo.m

```
%Auto encoder
load IMAGEDATA
LR=0.01;
%With 2500 neurons in the input layer, followed by
%1000,500,1000,2500. The output of the hidden neurons
%in the third layer is consider as the compressed
data
W1=randn(2501,1000)*0.1;
W2=randn(1001,500)*0.1;
W3=randn(501,1000)*0.1;
W4=randn(1001,2500)*0.1;
```

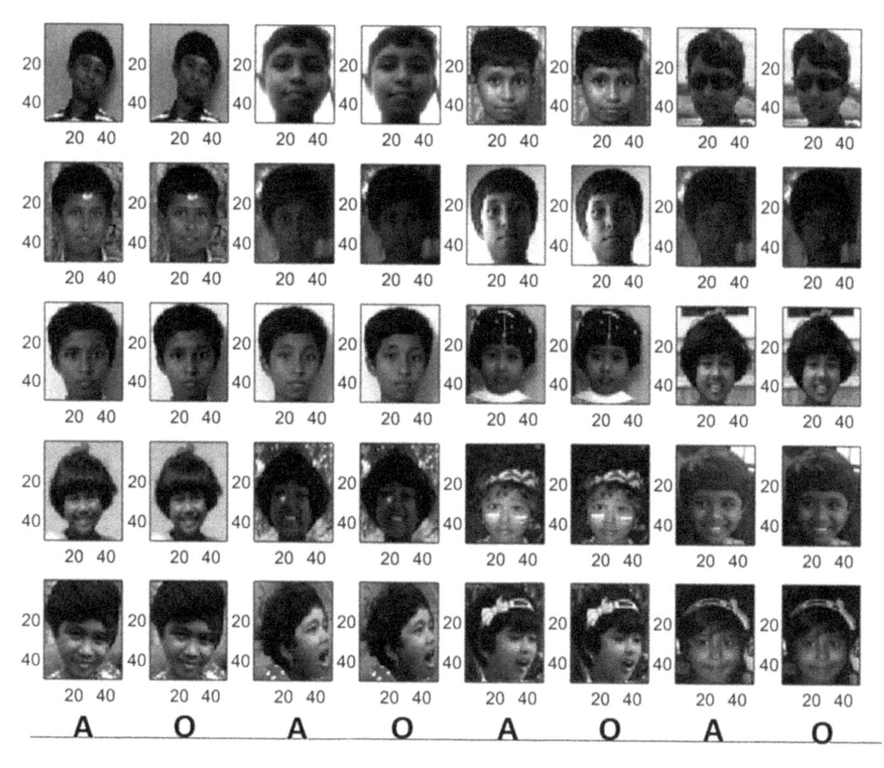

Fig. 5.36 Illustration of original images (O) and the reconstructed (A) images using autoencoder network

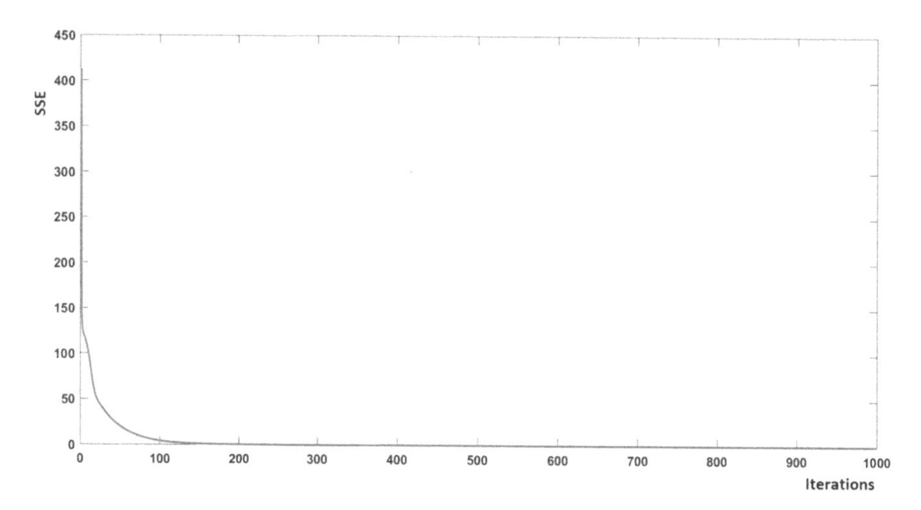

Fig. 5.37 Convergence obtained for autoencoder network

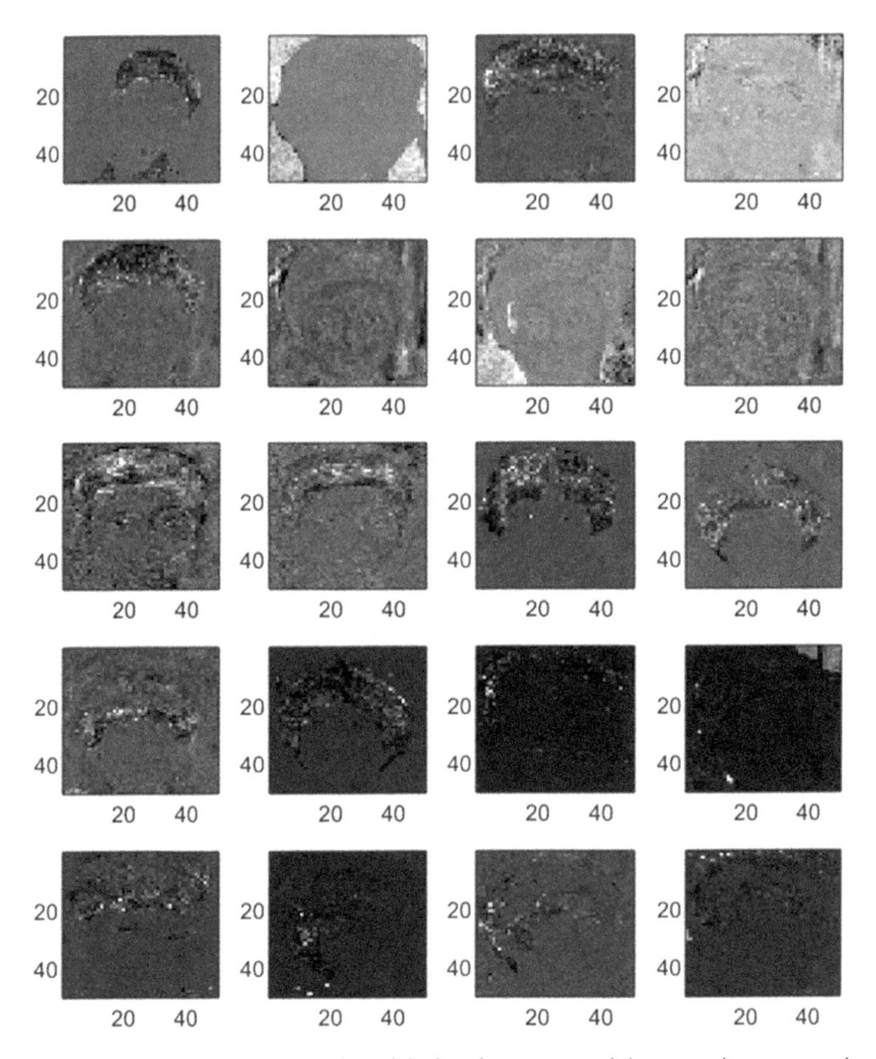

Fig. 5.38 The difference between the original and reconstructed images using autoencoder network

```
CLASS=[cell2mat(CLASS1');cell2mat(CLASS2')];
SSE=[];
for iteration=1:1:1000
    error=[];
for j=1:1:20
H1=logsig([CLASS(j,:) 1]*W1);
H2=logsig([H1 1]*W2);
H3=logsig([H2 1]*W3);
OUT=logsig([H3 1]*W4);
```

```
error=[error sum((CLASS(j,:)-OUT).^2)];
GRAD0=-(CLASS(j,:)-OUT).*OUT.*(1-OUT);
%Update weights
W4=W4-LR*[H3 1]'*GRAD0;
GRAD1=[];
for i=1:1:1000
W3(:,i)=W3(:,i)-[H2';1]*LR*sum((1-H3(i))*H3(i)*
(W4(i,:).*GRAD0));
GRAD1=[GRAD1;sum((1-H3(i))*H3(i)*W4(i,:).*GRAD0)];
end
GRAD2=[];
for i=1:1:500
W2(:,i)=W2(:,i)-[H1';1]*LR*sum((1-H2(i))*H2(i)*
W3(i,:)'.*GRAD1);
GRAD2=[GRAD2;sum((1-H2(i))*H2(i)*W3(i,:)'.*GRAD1)];
end
GRAD3=[];
for i=1:1:1000
W1(:,i)=W1(:,i)-[CLASS(j,:) 1]'*LR*sum((1-H1(i))*
H1(i)*W2(i,:)'.*GRAD2);
end
end
SSE=[SSE sum(sum(error.^2))];
end
plot(SSE/2500)
colormap(gray)
t=1;
t1=1;
for j=1:1:20
H1=logsig([CLASS(j,:) 1]*W1);
H2=logsig([H1 1]*W2);
H3=logsig([H2 1]*W3);
OUT=logsig([H3 1]*W4);
figure(2)
subplot(5,8,t)
t=t+1;
imagesc(reshape(OUT,[50 50]))
subplot(5,8,t)
t=t+1;
imagesc(reshape(CLASS(j,:),[50 50]))
figure(2)
subplot(5,4,t1)
imagesc(reshape(CLASS(j,:)-OUT,[50 50]))
t1=t1+1;
end
```

5.6.6 Recurrent Neural Network (RNN)

Consider the following patterns illustrated in Fig. 5.40. In this set, data in SEQUENCEDATA1 are arranged such that middle element is lesser than the first and the last element. Similarly, the data in SEQUENCEDATA2 are arranged such that middle element is greater than the first and the last element. If the first number is known, we understand that the second number is going to be lesser than the first element (SEQUENCEDATA1), and we can conclude that the second number is greater than the first and the last element (SEQUENCEDATA2). This additional information (hidden) is exploited by using RNN model to classify the data into one among the two category. The practical applications include the speech classification, image description, video description, language conversion, etc. The simple RNN model is shown in Fig. 5.25. In this model, weights are shared in all the three cascaded networks. Initially out_0 is assumed as 0. The input x_1 along with out_0 is given as input to the first network to obtain out_1. The out_1 along with x_2 is given as input to the second network, and the out_2 is obtained as the output. Finally, out_2 along with x_3 is given as input to the third network to obtain out_3 as the output. The gradients for the weight matrices are computed using the following equations. The gradient involves the product of several terms and hence there is the chance that gradient may vanish or may explode. To circumvent this, LSTM-based RNN is proposed and is discussed in the following section. It is noted that the proper combinations of sigmoidal function and tanh functions are used in training the network to avoid exploding or vanishing gradients. Illustrations for RNN are given in Figs. 5.39, 5.40, 5.41, and 5.42.

$$h_1^1 = logsig(out_0 w_{11} + x_1 w_{21})$$

$$h_2^1 = logsig(x_1 w_{22} + out_0 w_{12})$$

$$out_1 = logsig(h_1^1 v_1 + h_2^1 v_2 + v_3)$$

$$h_1^2 = logsig(out_1 w_{11} + x_2 w_{21})$$

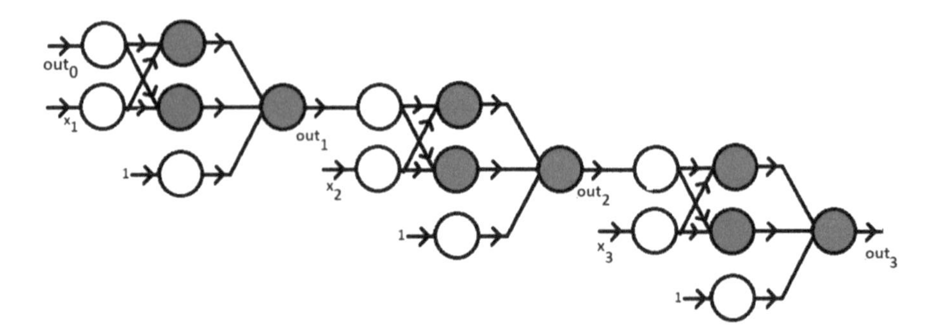

Fig. 5.39 Illustration of Recursive Neural Network

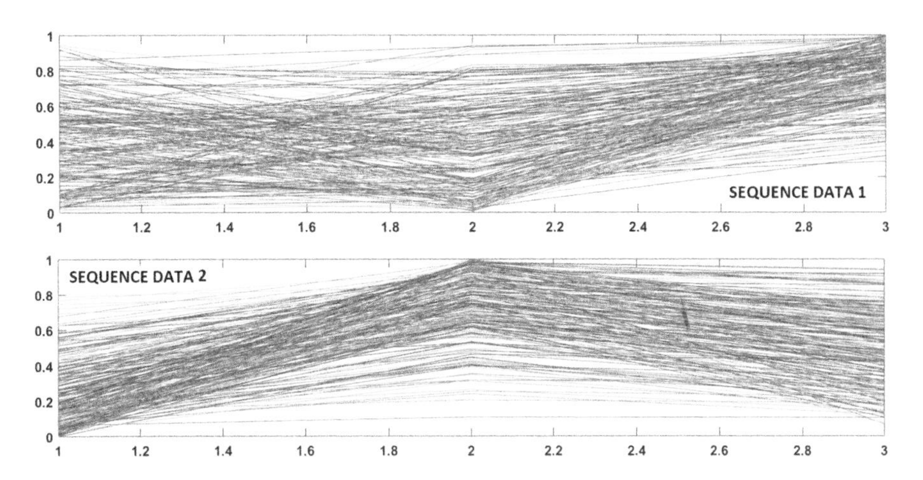

Fig. 5.40 Examples for the sequence data

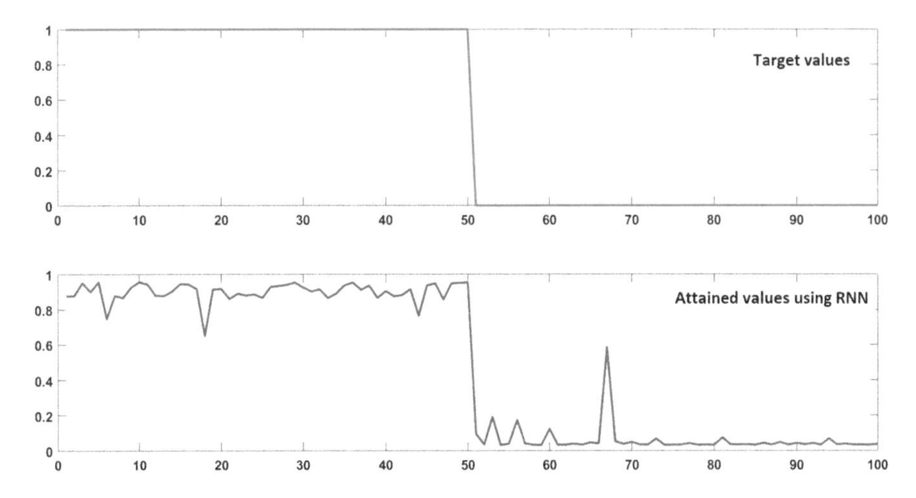

Fig. 5.41 Convergence achieved using RNN

$$h_2^2 = logsig(x_2 w_{22} + out_1 w_{12})$$

$$out_2 = logsig(h_1^2 v_1 + h_2^2 v_2 + v_3)$$

$$h_1^3 = logsig(out_2 w_{11} + x_3 w_{21})$$

$$h_2^3 = logsig(x_3 w_{22} + out_2 w_{12})$$

$$out_3 = logsig(h_1^3 v_1 + h_2^3 v_2 + v_3)$$

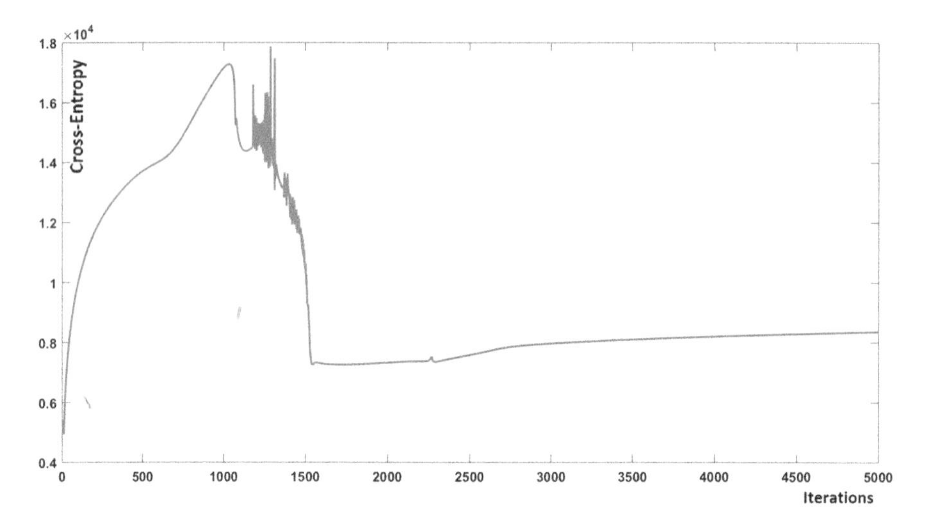

Fig. 5.42 Examples for the sequence data

RNNdemo.m

```
%Demonstrate RNN
clear all
close all
DATA1=[];
DATA2=[];
for i=1:1:100
t=sort(rand(1,3));
subplot(2,1,1)
DATA1=[DATA1;[t(2) t(1) t(3)]];
plot([t(2) t(1) t(3)])
hold on
subplot(2,1,2)
DATA2=[DATA2;[t(1) t(3) t(2)]];
plot([t(1) t(3) t(2)])
hold on
end
DATATRAIN=[DATA1(1:1:50,:); DATA2(1:1:50,:)];
DATATEST=[DATA1(51:1:100,:); DATA2(51:1:100,:)];
TARGETTRAIN=[ones(1,50) zeros(1,50)];
TARGETTEST=[ones(1,50) zeros(1,50)];
W=randn(2,2)*1;
V=randn(3,1)*1;
JCOL=[];
LR=0.05;
for j=1:1:5000
```

```
for i=1:1:size(DATATRAIN,1)
%Forward
out0=0;
h11=logsig(out0*W(1,1)+DATATRAIN(i,1)*W(2,1));
h21=logsig(out0*W(1,2)+DATATRAIN(i,1)*W(2,2));
out1=logsig(h11*V(1)+h21*V(2)+V(3));
h12=logsig(out1*W(1,1)+DATATRAIN(i,2)*W(2,1));
h22=logsig(out1*W(1,2)+DATATRAIN(i,2)*W(2,2));
out2=logsig(h12*V(1)+h22*V(2)+V(3));
h13=logsig(out2*W(1,1)+DATATRAIN(i,3)*W(2,1));
h23=logsig(out2*W(1,2)+DATATRAIN(i,3)*W(2,2));
out3=logsig(h13*V(1)+h23*V(2)+V(3));
%Backward
gradatout3=TARGETTRAIN(i)-out3;
V(1)=V(1)+LR*gradatout3*h13;
V(2)=V(1)+LR*gradatout3*h23;
V(3)=V(3)+LR*gradatout3;
W(1,1)=W(1,1)+LR*gradatout3*V(1)*h13*(1-h13)*out2;
W(1,2)=W(1,2)+LR*gradatout3*V(2)*h23*(1-h23)*out2;
W(2,1)=W(2,1)+LR*gradatout3*V(1)*h13*(1-h13)*
DATATRAIN(i,3);
W(2,2)=W(2,2)+LR*gradatout3*V(2)*h23*(1-h23)*
DATATRAIN(i,3);
gradatout2=gradatout3*(V(1)*h13*(1-h13)*W(1,1)+...
V(2)*h23*(1-h23)*W(1,2))*out2*(1-out2);
V(1)=V(1)+LR*gradatout2*h12;
V(2)=V(1)+LR*gradatout2*h22;
V(3)=V(3)+LR*gradatout2;
W(1,1)=W(1,1)+LR*gradatout2*V(1)*h12*(1-h12)*out1;
W(1,2)=W(1,2)+LR*gradatout2*V(2)*h22*(1-h22)*out1;
W(2,1)=W(2,1)+LR*gradatout2*V(1)*h12*(1-h12)*
DATATRAIN(i,2);
W(2,2)=W(2,2)+LR*gradatout2*V(2)*h22*(1-h22)*
DATATRAIN(i,2);
gradatout1=gradatout2*(V(1)*h12*(1-h12)*W(1,1)+...
V(2)*h22*(1-h22)*W(1,2))*out1*(1-out1);
V(1)=V(1)+LR*gradatout1*h11;
V(2)=V(1)+LR*gradatout1*h21;
V(3)=V(3)+LR*gradatout1;
W(1,1)=W(1,1)+LR*gradatout1*V(1)*h11*(1-h11)*out0;
W(1,2)=W(1,2)+LR*gradatout1*V(2)*h21*(1-h21)*out0;
W(2,1)=W(2,1)+LR*gradatout1*V(1)*h11*(1-h11)*
DATATRAIN(i,1);
W(2,2)=W(2,2)+LR*gradatout1*V(2)*h21*(1-h21)*
DATATRAIN(i,1);
```

```
end
EST=[];
for i=1:1:size(DATATRAIN,1)
%Forward
out0=0;
h11=logsig(out0*W(1,1)+DATATRAIN(i,1)*W(2,1));
h21=logsig(out0*W(1,2)+DATATRAIN(i,1)*W(2,2));
out1=logsig(h11*V(1)+h21*V(2)+V(3));
h12=logsig(out1*W(1,1)+DATATRAIN(i,2)*W(2,1));
h22=logsig(out1*W(1,2)+DATATRAIN(i,2)*W(2,2));
out2=logsig(h12*V(1)+h22*V(2)+V(3));
h13=logsig(out2*W(1,1)+DATATRAIN(i,3)*W(2,1));
h23=logsig(out2*W(1,2)+DATATRAIN(i,3)*W(2,2));
out3=logsig(h13*V(1)+h23*V(2)+V(3));
EST=[EST out3];
end
JCOL=[JCOL -1*sum(sum(TARGETTRAIN.*log(EST')))];
end
figure
plot(JCOL)
ESTTEST=[];
for i=1:1:size(DATATEST,1)
%Forward
out0=0;
h11=logsig(out0*W(1,1)+DATATEST(i,1)*W(2,1));
h21=logsig(out0*W(1,2)+DATATEST(i,1)*W(2,2));
out1=logsig(h11*V(1)+h21*V(2)+V(3));
h12=logsig(out1*W(1,1)+DATATEST(i,2)*W(2,1));
h22=logsig(out1*W(1,2)+DATATEST(i,2)*W(2,2));
out2=logsig(h12*V(1)+h22*V(2)+V(3));
h13=logsig(out2*W(1,1)+DATATEST(i,3)*W(2,1));
h23=logsig(out2*W(1,2)+DATATEST(i,3)*W(2,2));
out3=logsig(h13*V(1)+h23*V(2)+V(3));
ESTTEST=[ESTTEST out3];
end
figure
subplot(2,1,1)
plot(TARGETTEST,'r')
subplot(2,1,2)
plot(ESTTEST,'b')
```

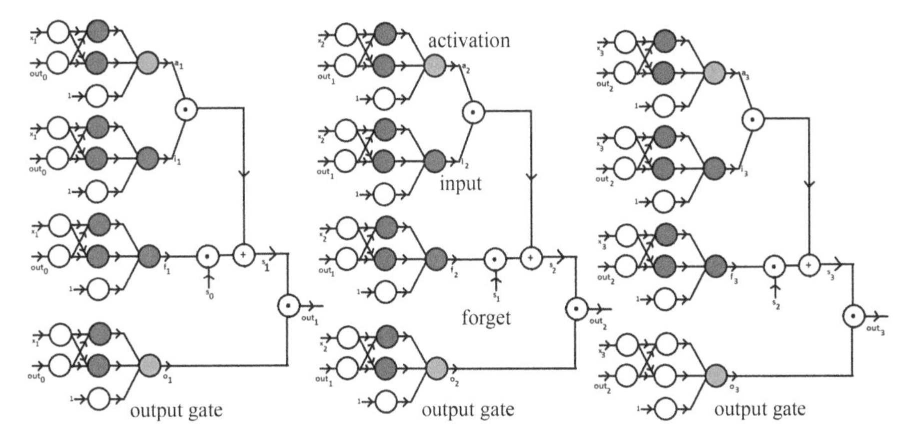

Fig. 5.43 Illustration of Long–Short Term Memory

5.6.7 *Long–Short Term Memory Network*

To circumvent the problem of vanishing gradient and exploding gradient in Recurrent Neural Network, Long–Short Term Memory Network was proposed. The set of gradients used to update the weights are summarized below. The link between the ith neuron of the first layer and the jth neuron of the second layer is represented as w_{ij}^{g}. The link between the ith neuron in the second layer and the jth neuron in the third layer is represented as v_{ij}^{g}. In this g indicates the gate identity that takes either a, f, i, or o respectively for the (a) activation gate, (b) forget gate, (c) input gate, and (d) output gate. The identical weights are used for the corresponding gate for three consecutive inputs x_1, x_2, and x_3. The relationship between various variables is summarized. These variables are used to obtain the gradient equations and hence training is achieved. The illustration of Long–Short Term Memory is given in Fig. 5.43.

$$out_3 = s_3 + o_3, out_2 = s_2 + o_2, out_1 = s_1 + o_1$$

$$s_3 = (f_3 \cdot s_2) + (a_3 \cdot i_3), s_2 = (f_2 \cdot s_1) + (a_2 \cdot i_2), s_1 = (f_1 \cdot s_0) + (a_1 \cdot i_1)$$

$$o_3 = tanh(h_{11}^{o} v_{11}^{o} + h_{12}^{o} v_{21}^{o} + v_{31}^{o}), o_2 = tanh(h_{21}^{o} v_{11}^{o} + h_{22}^{o} v_{21}^{o} + v_{31}^{o}),$$

$$o_1 = tanh(h_{31}^{o} v_{11}^{o} + h_{32}^{o} v_{21}^{o} + v_{31}^{o})$$

$$f_3 = logsig(h_{11}^{f} v_{11}^{f} + h_{12}^{f} v_{21}^{f} + v_{31}^{f}), f_2 = logsig(h_{21}^{f} v_{11}^{f} + h_{22}^{f} v_{21}^{f} + v_{31}^{f}),$$

$$f_1 = logsig(h_{31}^{f} v_{11}^{f} + h_{32}^{f} v_{21}^{f} + v_{31}^{f})$$

$$i_3 = logsig(h_{11}^{i} v_{11}^{i} + h_{12}^{i} v_{21}^{i} + v_{31}^{i}), i_2 = logsig(h_{21}^{i} v_{11}^{i} + h_{22}^{i} v_{21}^{i} + v_{31}^{i}),$$

$$i_1 = logsig(h_{31}^{i} v_{11}^{i} + h_{32}^{i} v_{21}^{i} + v_{31}^{i})$$

$$a_3 = tansig(h^a_{11} v^i_{11} + h^a_{12} v^i_{21} + v^a_{31}), a_2 = tansig(h^a_{21} v^i_{11} + h^a_{22} v^i_{21} + v^a_{31}),$$
$$a_1 = tansig(h^a_{31} v^i_{11} + h^a_{32} v^i_{21} + v^a_{31})$$
$$h^o_{11} = logsig(x_1 w^o_{11} + out_0 w^o_{21}), h^o_{12} = logsig(x_1 w^o_{11} + out_0 w^o_{21}),$$
$$h^o_{21} = logsig(x_2 w^o_{11} + out_1 w^o_{21})$$
$$h^o_{22} = logsig(x_2 w^o_{11} + out_1 w^o_{21}), h^o_{31} = logsig(x_3 w^o_{11} + out_2 w^o_{21}),$$
$$h^o_{32} = logsig(x_3 w^o_{11} + out_2 w^o_{21})$$
$$h^f_{11} = logsig(x_1 w^o_{11} + out_0 w^o_{21}), h^f_{12} = logsig(x_1 w^o_{11} + out_0 w^o_{21}),$$
$$h^f_{21} = logsig(x_2 w^o_{11} + out_1 w^o_{21})$$
$$h^f_{22} = logsig(x_2 w^o_{11} + out_1 w^o_{21}), h^f_{31} = logsig(x_3 w^o_{11} + out_2 w^o_{21}),$$
$$h^f_{32} = logsig(x_3 w^o_{11} + out_2 w^o_{21})$$
$$h^i_{11} = logsig(x_1 w^i_{11} + out_0 w^i_{21}), h^i_{12} = logsig(x_1 w^i_{11} + out_0 w^i_{21}),$$
$$h^i_{21} = logsig(x_2 w^i_{11} + out_1 w^i_{21})$$
$$h^i_{22} = logsig(x_2 w^i_{11} + out_1 w^i_{21}), h^i_{31} = logsig(x_3 w^i_{11} + out_2 w^i_{21}),$$
$$h^i_{32} = logsig(x_3 w^i_{11} + out_2 w^i_{21})$$
$$h^a_{11} = logsig(x_1 w^a_{11} + out_0 w^a_{21}), h^a_{12} = logsig(x_1 w^a_{11} + out_0 w^a_{21}),$$
$$h^a_{21} = logsig(x_2 w^a_{11} + out_1 w^a_{21})$$
$$h^a_{22} = logsig(x_2 w^a_{11} + out_1 w^a_{21}), h^a_{31} = logsig(x_3 w^a_{11} + out_2 w^a_{21}),$$
$$h^a_{32} = logsig(x_3 w^a_{11} + out_2 w^a_{21})$$

Chapter 6
Statistical Test in Pattern Recognition

6.1 Statistical Test to Validate the Performance of the Classifiers

Let x_i be the percentage of success obtained using the constructed classifier in the ith (with $i = 1 \cdots M$ and $j = 1 \cdots N$) attempt. Each result is viewed as the outcome of the random variable X_i with $E(X_i) = \mu_p(mean)$. Consider the problem of providing the statistical evidence to claim the performance of the classifier (in terms of detection rate) greater than $p\%$ with following cases.

1. Case 1: X_i is assumed as normal distributed with known variance σ^2

$$Z = \frac{1}{N} \sum_{i=1}^{i=N} \frac{(X_i - p)}{\sqrt{\frac{\sigma^2}{N}}} \tag{6.1}$$

   Now the requirement is to check whether the classifier detection rate is greater than $p\%$. We formulate the null hypothesis such that the classifier detection rate is lesser than $p\%$, i.e., $H_0 : \mu_z < 0$. To check with 95% confident level, we choose z_α such that $p(Z \leq z_\alpha) = 0.95$ (using normal distribution) and reject the null hypothesis if the typical outcome z_{test} computed using the observed outcomes as $z_{test} = \frac{1}{N} \sum_{i=1}^{i=N} \frac{(x_i - p)}{\sqrt{\frac{\sigma^2}{N}}}$ is greater than z_α. That is, if $z_{test} \geq z_\alpha$, we conclude that the detection rate of the constructed classifier is greater than $p\%$ with 95% confident level.
2. Case 2: $N > 30$, the random variable X_i is assumed as normal distributed (sample size is large enough to apply the central limit theorem) and the actual variance of the random variable X_i is not known.

© Springer Nature Switzerland AG 2020

E. S. Gopi, *Pattern Recognition and Computational Intelligence Techniques Using Matlab*, Transactions on Computational Science and Computational Intelligence,
https://doi.org/10.1007/978-3-030-22273-4_6

$$Z = \frac{1}{N} \sum_{i=1}^{i=N} \frac{(X_i - p)}{\sqrt{\frac{s}{N}}} \tag{6.2}$$

where s is computed as $s^2 = \frac{1}{(N-1)} \sum_{j=1}^{j=N} (x_j - p)^2$ where s^2 is the estimate of the variance of the random variable X_i. It is noted in the expression actual outcomes (x_j) are used. In this case, Z follows the normal distribution.

Now the requirement is to check whether the classifier detection rate is greater than $p\%$. We formulate the null hypothesis such that the classifier detection rate is lesser than $p\%$, i.e., $H_0 : \mu_z < 0$. To check with 95% confident level, we choose z_α such that $p(Z \leq z_\alpha) = 0.95$ (using normal distribution) and reject the null hypothesis if the typical outcome z_{test} computed using the observed outcomes as $z_{test} = \frac{1}{N} \sum_{i=1}^{i=N} \frac{(x_i - p)}{\sqrt{\frac{s}{N}}}$ is greater than z_α. That is, if $z_{test} \geq z_\alpha$, we conclude that the detection rate of the constructed classifier is greater than $p\%$ with 95% confident level.

3. Case 3: $N < 30$, the random variable X_i is assumed as normal distributed and the actual variance of the random variable X_i is not known.

$$Z = \frac{1}{N} \sum_{i=1}^{i=N} \frac{(X_i - p)}{\sqrt{\frac{S^2}{N}}} \tag{6.3}$$

where s is computed as $S^2 = \frac{1}{(N-1)} \sum_{j=1}^{j=N} (X_{pj} - \mu_p)^2$.

In this case Z follows Student's t distribution with order $N - 1$.

Now the requirement is to check whether the classifier detection rate is greater than $p\%$. We formulate the null hypothesis such that the classifier detection rate is lesser than $p\%$, i.e., $H_0 : \mu_z < 0$. To check with 95% confident level, we choose z_α such that $p(Z \leq z_\alpha) = 0.95$ (using t-distribution) and reject the null hypothesis if the typical outcome z_{test} computed using the observed outcomes as $z_{test} = \frac{1}{N} \sum_{i=1}^{i=N} \frac{(x_i - p)}{\sqrt{\frac{s^2}{N}}}$ (where $s^2 = \frac{1}{(N-1)} \sum_{j=1}^{j=N} (x_j - p)^2$) is greater than z_α. If $z_{test} \geq z_\alpha$, we conclude that the detection rate of the constructed classifier is greater than $p\%$ with 95% confident level.

meantest.m

```
%Mean test without known variance
%To test whether the average Percentage of Success
(POS)
%obtained using the method A is the outcome
%of the random variable with mean 90
A=[92.7526 90.138 88.0929 89.6350 89.1519 ...
   89.2352 88.8723 90.0782 92.1066 89.2842 ...
   89.7195 91.1665 91.2128 90.4855 91.0260 ...
   90.8707 89.6182 90.4289 89.7009 89.1001];
```

```
%Obtain the outcome of the random variable T (ttest)
with
%zero mean and variance 1 as follows.
M=mean(A);
V=var(A);
ttest=(M-90)/sqrt(V/20);
%As the number of outcomes is lesser than $30$ and Z
%follows $t$ distribution with order 19
c=cumsum((pdf('t',-5:0.1:5,19)/10));
t=-5:0.1:5;
plot(t,c);
%Locating 0.05 significant level (95% confident level)
%is the value at which the cumulative distribution is
%0.05 and 0.95
[p1,q1]=find((c-0.05)>0);
[p2,q2]=find((c-0.95)>0);
hold on
plot([t(q1(1)) t(q1(1))],[0 c(q1(1))],'-r*')
hold on
plot([t(q2(1)) t(q2(1))],[0 c(q2(1))],'-r*')
%The Null Hypotheis is structured such that the
%mean is equal to 90
%We reject the Hypotheis if ttest is greater than
t(q1(1))
%or less than t(q2(1))
if((ttest<t(q1(1)))|(ttest>t(q2(1))))
    disp('Reject the Null Hypothesis (i.e., mean is
    equal to 90)')
else
    disp('Accept the Null Hypothesis (i.e., mean is
    equal to 90)')
end
```

6.2 Statistical Test to Compare the Performance of the Classifiers

Let x_{ij} (with $i = 1 \cdots M$ and $j = 1 \cdots N$) be the detection rate (in percentage) obtained by the ith classifier in the jth trial. We treat the actual values obtained as the outcome of the corresponding random variables X_{ij}. To compare the classifier p and q, we construct the random variable Z_{pq} that depends upon various cases as follows:

1. Case 1: X_{pj} is normal distributed with known variance σ_p^2.

$$Z_{pq} = \frac{\frac{1}{N}\sum_{j=1}^{j=N}(X_{pj} - X_{qj})}{\sqrt{\frac{\sigma_p^2+\sigma_q^2}{N}}} \tag{6.4}$$

   In this case, Z_{pq} follows normal distribution.

2. Case 2: $N > 30$, the random variable X_{pj} is assumed as normal distributed (sample size is large enough to apply the central limit theorem) and the actual variance of the random variable X_{pj} is not known.

$$Z_{pq} = \frac{\frac{1}{N}\sum_{j=1}^{j=N}(X_{pj} - X_{qj})}{\sqrt{\frac{s_p^2+s_q^2}{N}}} \tag{6.5}$$

   where $s_p^2 = \frac{1}{(N-1)}\sum_{j=1}^{j=N}(x_{pj} - \mu_p)^2$ and $s_q^2 = \frac{1}{(N-1)}\sum_{j=1}^{j=N}(x_{qj} - \mu_q)^2$ are the estimate of the variances of X_{pj} and X_{qj}, respectively. It is also observed that $\mu_{pq} = E(Z_{pq}) = \frac{\mu_p-\mu_q}{\sqrt{\frac{s_p^2+s_q^2}{N}}}$ and $var(Z_{pq}) = 1$, where μ_p and μ_q are the estimate of the mean of the random variables X_{pj} and X_{qj} respectively (for all values of j).

3. Case 3: $N < 30$, the random variable X_{pj} is normal distributed with unknown variance.

$$Z_{pq} = \frac{\frac{1}{N}\sum_{j=1}^{j=N}(X_{pj} - X_{qj})}{\sqrt{\frac{2S^2}{N}}} \tag{6.6}$$

   where $S^2 = \frac{S_p^2(N-1)+S_q^2(N-1)}{2N-2}$, $S_p^2 = \frac{1}{(N-1)}\sum_{j=1}^{j=N}(X_{pj} - \mu_p)^2$, and $S_q^2 = \frac{1}{(N-1)}\sum_{j=1}^{j=N}(X_{qj} - \mu_q)^2$ are the random variables associated with p^{th} and the qth classifier, respectively. In this case, Z_{pq} follows Student's *t* distributed with order $2N - 1$.

Now the requirement is to check whether there is significant difference in the performance of the classifier p and q using the actual observed outcomes x_{ij}. We formulate the null hypothesis such that there is no significant change in the performance of the p and q, i.e., $H_0 : \mu_{pq} = 0$. To check with 95% confident level, we choose z_α such that $p(-z_\alpha \leq Z_{pq} \leq z_\alpha) = 0.95$ (using normal distribution) and reject the null hypothesis if the typical outcome z_{test} computed using the observed outcomes as $z_{test} = \frac{\frac{1}{N}\sum_{j=1}^{j=N}(x_{pj}-x_{qj})}{\sqrt{\frac{\sigma_p^2+\sigma_q^2}{N}}}$ (case 1), $z_{test} = \frac{\frac{1}{N}\sum_{j=1}^{j=N}(x_{pj}-x_{qj})}{\sqrt{\frac{s_p^2+s_q^2}{N}}}$ (case 2) is greater than z_α. Similarly, for the case 3, we choose z_α such that $p(-z_\alpha \leq Z_{pq} \leq z_\alpha) = 0.95$ (using Student's *t* distribution with order $2N - 1$) and reject the null hypothesis if the typical outcome z_{test} computed using the observed outcomes $Z_{test} = \frac{\frac{1}{N}\sum_{j=1}^{j=N}(x_{pj}-x_{qj})}{\sqrt{\frac{s^2}{N}}}$ (case 3) is greater than z_α, where $s^2 = \frac{s_p^2(N-1)+s_q^2(N-1)}{2N-2}$, $s_p^2 = \frac{1}{(N-1)}\sum_{j=1}^{j=N}(x_{pj}-\mu_p)^2$, and $s_q^2 = \frac{1}{(N-1)}\sum_{j=1}^{j=N}(x_{qj}-\mu_q)^2$.

meantestcompare.m

```
%Mean test without known variance
%To test whether the average Percentage of Success
(POS)
%obtained using the method A and B are identical
A= [95.0771 90.7768 87.7402 89.4356 90.9015 90.3947 ...
    90.0049 90.4369 91.1301 90.1538 89.2414 89.8198 ...
    89.7922 90.8967 90.4123 90.5475 90.1478 89.6377 ...
    90.0611 90.2167];

B=[89.2501 88.0406 87.7227 90.1286 87.9161 88.0061 ...
   91.9278 92.4740 97.7052 90.9563 90.1844 88.1844 ...
   89.3827 89.8402 84.4478 89.1862 92.3703 88.1951 ...
   91.5629 93.1352];
%Obtain the outcome of the random variable T (ttest)
with
%zero mean and variance 1 as follows.
%Pooled variance
s=(var(A)*(19)+var(B)*(19))/38;
ttest=(mean(A)-mean(B))/sqrt(s/20);
%As the number of outcomes is lesser than $30$ and Z
%follows $t$ distribution with order 20+20-2=38
c=cumsum((pdf('t',-5:0.1:5,38)/10));
t=-5:0.1:5;
plot(t,c);
%Locating 0.05 significant level (95% confident level)
%is the value at which the cumulative distribution is
%0.05 and 0.95
[p1,q1]=find((c-0.05)>0);
[p2,q2]=find((c-0.95)>0);
hold on
plot([t(q1(1)) t(q1(1))],[0 c(q1(1))],'-r*')
hold on
plot([t(q2(1)) t(q2(1))],[0 c(q2(1))],'-r*')
plot(ttest,0,'bo')
%The Null Hypotheis is structured such that the
%mean is equal to 90
%We reject the Hypotheis if ttest is greater than
t(q1(1))
```

```
%or less than t(q2(1))
%Formulate the Null Hypothesis
disp('Null Hypothesis: Both A and B are the outcomes
of the ...
normal distribution with identical means')
if((ttest<t(q1(1)))|(ttest>t(q2(1))))
    disp('Reject the Null Hypothesis with 95 percent
    confident')
else
    disp('Accept the Null Hypothesis with 95 percent
    confident')
end
```

6.2.1 Summary on the Mean Test

- If $N > 30$, we can still use t-distribution as described in Case 3.
- It is observed that $\frac{Z}{\frac{\psi^2}{N}}$ follows t distribution with order $N - 1$.
- Suppose that the number of outcomes of the individual classifier p (N_1) is different from the outcomes of the classifier q (N_2), then the following modifications are incorporated for the cases described in Sect. 6.1.

1. Case 1: $Z_{pq} = \frac{\frac{1}{N}\sum_{j=1}^{j=N}(X_{pj}-X_{qj})}{\sqrt{\frac{\sigma_p^2}{N_1}+\frac{\sigma_q^2}{N_2}}}$
2. Case 2: $Z_{pq} = \frac{\frac{1}{N}\sum_{j=1}^{j=N}(X_{pj}-X_{qj})}{\sqrt{\frac{s_p^2}{N_1}+\frac{s_q^2}{N_2}}}$, where $s_p^2 = \frac{1}{(N1-1)}\sum_{j=1}^{j=N1}(x_{pj} - \mu_p)^2$ and $s_q^2 = \frac{1}{(N2-1)}\sum_{j=1}^{j=N2}(x_{qj} - \mu_q)^2$ are the estimate of the variances of X_{pj} and X_{qj}, respectively. It is also observed that $\mu_{pq} = E(Z_{pq}) = \frac{\mu_p-\mu_q}{\sqrt{\frac{s_p^2+s_q^2}{N}}}$ and $var(Z_{pq}) = 1$, where μ_p and μ_q are the estimate of the mean of the random variables X_{pj} and X_{qj} respectively (for all values of j).
3. Case 3: $Z_{pq} = \frac{\frac{1}{N}\sum_{j=1}^{j=N}(X_{pj}-X_{qj})}{\sqrt{\frac{S^2}{N1}+\frac{S^2}{N2}}}$, with $S^2 = \frac{S_p^2(N1-1)+S_q^2(N2-1)}{N1+N2-2}$, $S_p^2 = \frac{1}{(N1-1)}\sum_{j=1}^{j=N1}(X_{pj} - \mu_p)^2$, and $S_q^2 = \frac{1}{(N2-1)}\sum_{j=1}^{j=N2}(X_{qj} - \mu_q)^2$ are the random variables associated with p^{th} and the qth classifier, respectively. The illustration of mean test without known variance is illustrated in Figs. 6.1, and 6.2, respectively.

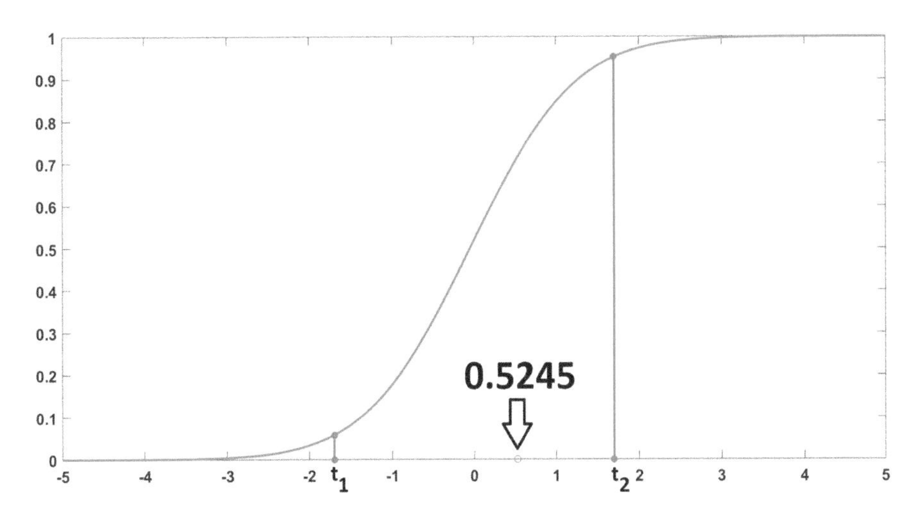

Fig. 6.1 Illustration of t-test with unknown variance to test the mean performance of the typical classifier (refer meantest.m)

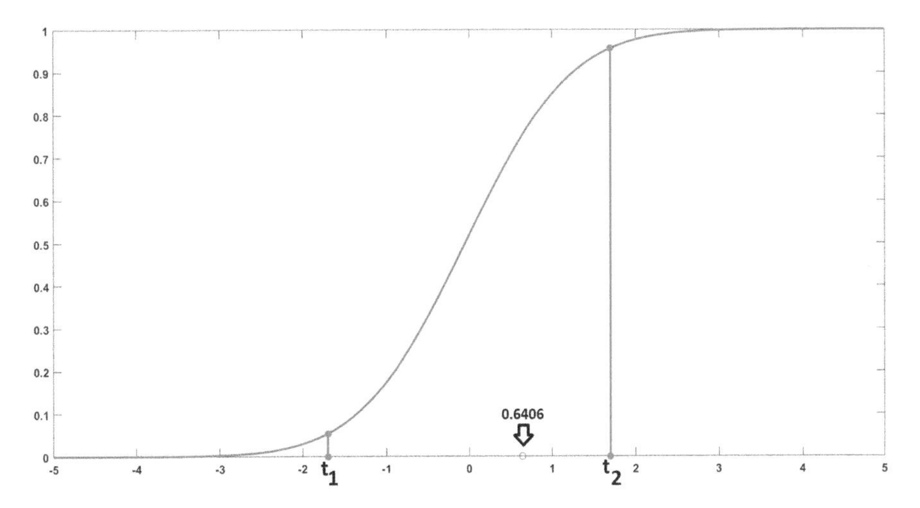

Fig. 6.2 Illustration of t-test with unknown variance to compare the performance of two classifiers (refer meantestcompare.m)

6.3 Binomial Test

Suppose we want to compare the performance of two methods A and B. Let us assume that we have performed 30 experiments, and 20 times method A has performed better than method B. The requirement is to check whether 75% of the time the percentage of method A is better than method B. That is, is that the outcome 20 out of 30 is the statistical evidence in declaring that it is the outcome

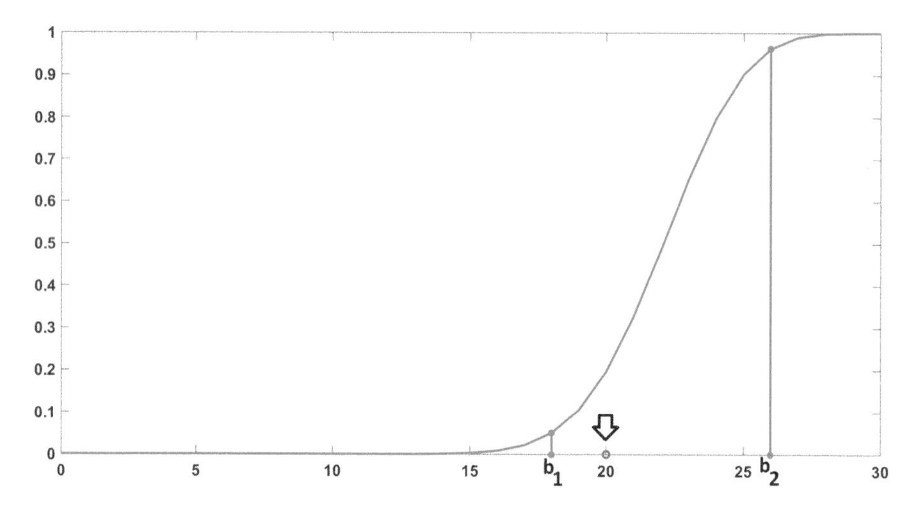

Fig. 6.3 Illustration of binomial distribution for percentage of success test

from the Binomial distribution with success rate $p = 0.75$. Consider the cumulative distribution of the binomial distribution with probability of success as $p = 0.75$ (refer Fig. 6.3). If the actual success out of 30 is greater than b_2 or less than b_1 (refer Fig. 6.3), then we declare that the success rate is not equal to 75% with 95% of confident.

binomialtest.m

```
%Binomial test
%Suppose the observation of 20 success out of 30
attempts
%is the statistical evidence (with 95% confident
level)
%that the outcome is from
%Binomial distribution with success rate 0.75
v=[];
for n=0:1:30
v=[v (factorial(30)/(factorial(30-n)*factorial(n)))*
((0.75)^n)*((0.25)^(30-n))];
end
t=0:1:30;
c=cumsum(v);
figure
plot(t,c,'-r')
%Locating 0.05 significant level (95% confident level)
%is the value at which the cumulative distribution is
%0.05 and 0.95
[p1,q1]=find((c-0.05)>0);
```

```
[p2,q2]=find((c-0.95)>0);
hold on
plot([t(q1(1)) t(q1(1))],[0 c(q1(1))],'-r*')
hold on
plot([t(q2(1)) t(q2(1))],[0 c(q2(1))],'-r*')
plot(20,0,'bo')
%The Null Hypotheis is structured such that the
%probabilty of sucess is 0.75
%We reject the Hypotheis if 20 (observation) is
greater than t(q1(1))
%or less than t(q2(1))
if((20<t(q1(1)))|(20>t(q2(1))))
    disp('Reject the Null Hypothesis ...
(i.e., probability of success is equal to 0.75)')
else
    disp('Accept the Null Hypothesis ...
(i.e., probability of success is equal to 0.75)')
end
```

6.4 Variance Test (Chi-Square test)

The percentage of success (POS) obtained using the typical outcome is given. Let the number of outcomes be N. We need to statistically test whether or not variance of the performance (in terms of POS) is equal to 0.1. The variance is estimated using the outcome as s^2. Compute $c = \frac{(N-1)s^2}{\sigma^2}$, where c is assumed as the outcome of the random variable that follows chi-square with order $(N-1)$. If $s_2 < c < s_1$, we declare that the variance of the performance is not equal to 0.1 with 95% confident level. The illustration of variance test is given in Fig. 6.4.

variancetest.m

```
%Statistical test for variance.
%To check the outcome of the random variable X
%that are from Gaussian distributed with the
particular
%variance
DATA=[89.6790 89.8510 90.0433 89.9077 90.0954 ...
     90.1265 89.7059 89.9441 89.3258 90.3622 ...
     89.8011 89.6193 89.9197 89.5482 89.9934 ...
     89.8227 90.6887 90.3600 89.2104 90.1396]
v=var(DATA);
disp('Hypothesis test for variance test')
disp('Null Hypothesis:variance is equal to 0.1')
s=((15-1)*v)/(0.1);
%s follows chi-square distribution with order 15-1
c=cumsum((pdf('chi',0:0.1:50,19)/10));
```

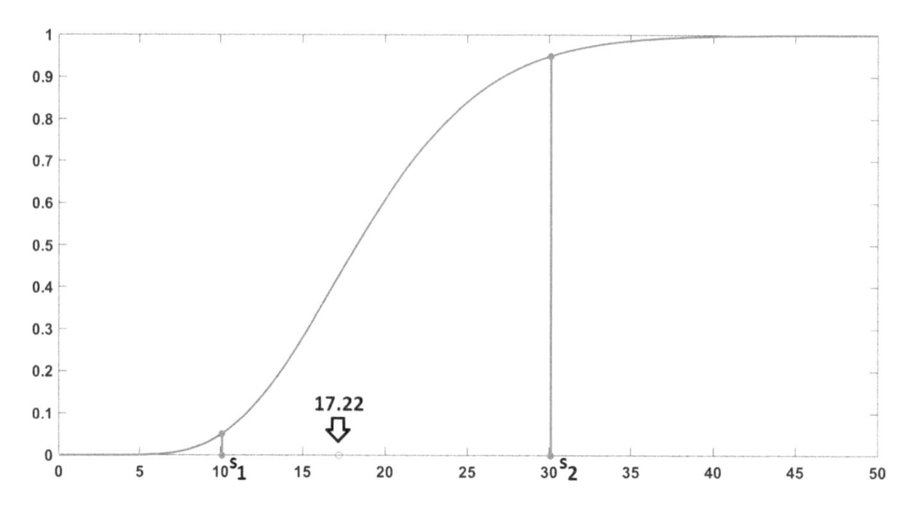

Fig. 6.4 Illustration of chi-square distribution for variance test

```
t=0:0.1:50;
plot(t,c);
%Locating 0.05 significant level (95% confident level)
%is the value at which the cumulative distribution is
%0.05 and 0.95
[p1,q1]=find((c-0.05)>0);
[p2,q2]=find((c-0.95)>0);
hold on
plot([t(q1(1)) t(q1(1))],[0 c(q1(1))],'-r*')
hold on
plot([t(q2(1)) t(q2(1))],[0 c(q2(1))],'-r*')
plot(s,0,'bo')
%We reject the Hypotheis if s is greater than t(q1(1))
%or less than t(q2(1))
if(s>t(q2(1)))
    disp('Reject the Null Hypothesis (i.e., var is
    equal to 0.1)')
else
    disp('Accept the Null Hypothesis (i.e., var is
    equal to 0.1)')
end
```

6.5 Proportion Test

Let the performance obtained using various methods using five different data sets are given in the matrix **M** (3×5 table). Each row indicates different method, and each column indicates different dataset. The requirement is to check whether the performance of the three methods is independent (proportion test) of data set. Let f_i be the summation of the ith row of the matrix and n_j is the summation of the jth column. The statistical test parameter t is computed as follows:

$$e(i, j) = \frac{f_i n_j}{n}$$

$$t = \sum_{i=1}^{i=3} \sum_{j=1}^{j=5} \left(\frac{(m(i, j) - e(i, j))^2}{e(i, j)}\right)$$

The t thus obtained is assumed as the outcome of the t-distribution with order $(3-1)(5-1)$. The illustration of proportion test is given in Fig. 6.5.

proportiontest.m

```
%Proportion test
%Each row of the matrix is the performance of the
%three methods for classification
%Each column indicate the particular dataset.
disp('Null Hypothesis: Performance of three methods ...
are independent of dataset')
DATA=[90    90    90    88    90;
    92    90    91    91    91;
    91    89    91    89    90]
F=sum(DATA);
N=sum(DATA');
for i=1:1:3
    for j=1:1:5
        e(i,j)=N(i)*F(j)/sum(N);
    end
end
%Obtain the outcome of the random variable t
%which is t distributed with order(3-1)*(5-1)
testt=0;
for i=1:1:3
    for j=1:1:5
testt=testt+((DATA(i,j)-e(i,j))^2)/(e(i,j));
    end
end
c=cumsum((pdf('t',-5:0.1:5,8)/10));
t=-5:0.1:5;
plot(t,c);
%Locating 0.05 significant level (95% confident level)
%is the value at which the cumulative distribution is
%0.05 and 0.95
[p1,q1]=find((c-0.05)>0);
[p2,q2]=find((c-0.95)>0);
```

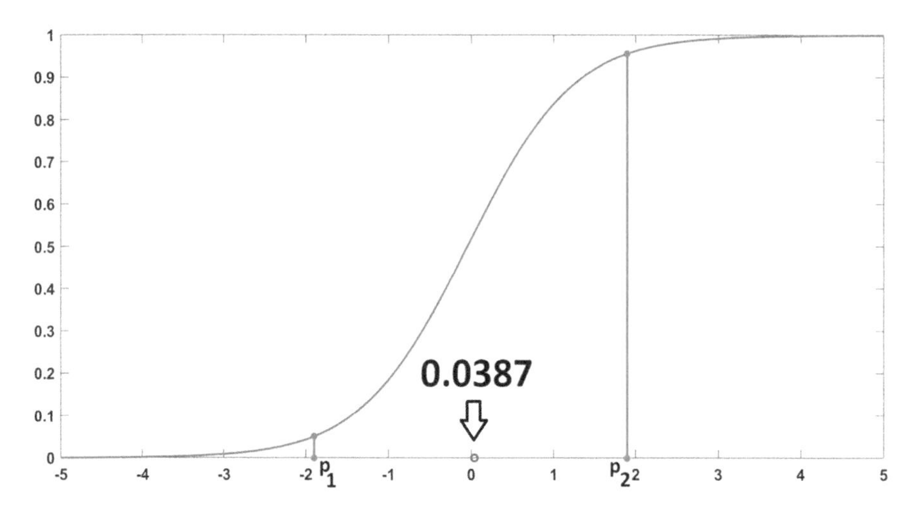

Fig. 6.5 Illustration of t-test for proportion test

```
hold on
plot([t(q1(1)) t(q1(1))],[0 c(q1(1))],'-r*')
hold on
plot([t(q2(1)) t(q2(1))],[0 c(q2(1))],'-r*')
plot(testt,0,'bo')
%The Null Hypotheis is structured such that the
%mean is equal to 90
%We reject the Hypotheis if ttest is gtreater than
t(q1(1))
%or less than t(q2(1))
%Formulate the Null Hypothesis
if((testt<t(q1(1)))|(testt>t(q2(1))))
    disp('Reject the Null Hypothesis with 95 percent
    confident')
else
    disp('Accept the Null Hypothesis with 95 percent
    confident')
end
```

6.6 Analysis of Variance (ANOVA) (F-Test)

Let the outcomes of the individual classes be given. We would like to test whether the mean of the individual classes are identical, then ANOVA (F-test) is used for the same. Let $\mathbf{x_{ij}}$ be the jth vector in the ith class. The between-class scatter variance and the within-class scatter variance are computed as follows:

$$S_W = \sum_{i=1}^{i=k} \sum_{j=1}^{j=n_i} (X_{ij} - c_i)^2 \tag{6.7}$$

$$S_B = \sum_{i=1}^{i=k} n_i (c_i - c)^2 \tag{6.8}$$

$F = \frac{\frac{S_B}{k-1}}{\frac{S_W}{N-k}}$ follows the F-distribution with order $(k-1, N-k)$, if the population mean of the individual class is identical. It is noted that the ANOVA test assumes that the within-class variance and the between-class variance are identical (σ^2). It is noted that the unbiased estimate of the between-class variance and the within-class variance are obtained as $\frac{S_B}{k-1}$ and $\frac{S_W}{N-k}$ respectively, i.e., $E\left(\frac{S_B}{k-1}\right) = \sigma^2$ and $E\left(\frac{S_W}{N-k}\right) = \sigma^2$. The expectation of the random variable F is 1. We formulate the null hypothesis as mean of all the individual classes are identical, and the alternative hypothesis is formulated such that at least two classes are not having identical mean. The illustration of the F-test is given in Figs. 6.6 and 6.8 corresponding to 10 class data illustrated, respectively, in Figs. 6.9 and 6.7.

Fscore.m

```
c=10;
 n=round(rand(1,c)*45)+5;
 AM=round(rand(1,c)*39)+1;
 for i=1:1:c
 Group{i}=randn(1,n(i))*500+AM(i);
```

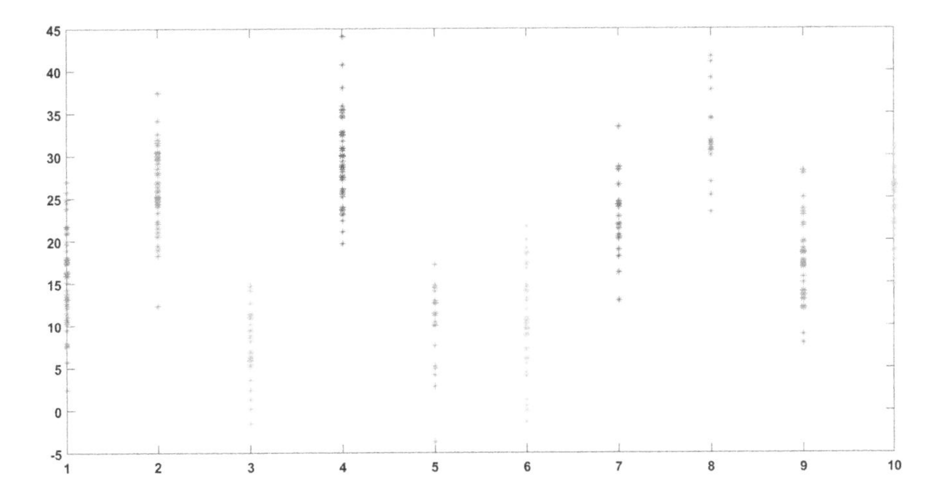

Fig. 6.6 Data under mean test using F-test (with distinct mean) (Data 1)

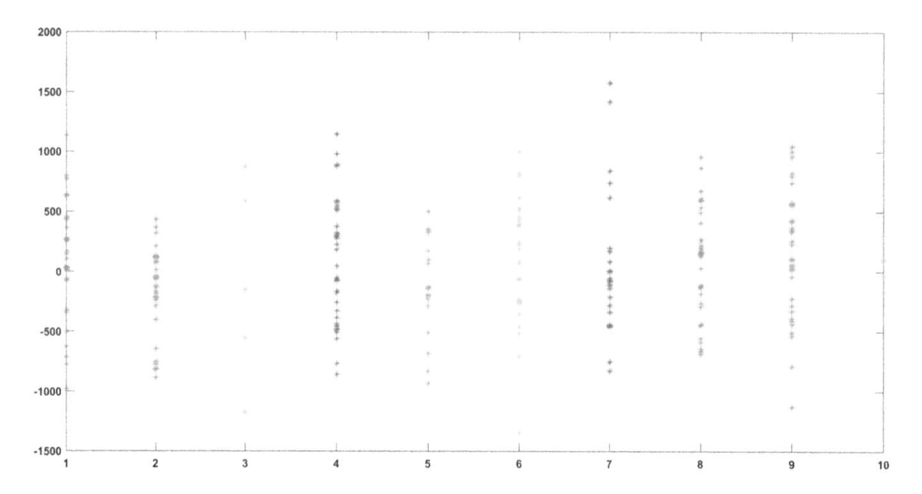

Fig. 6.7 Data under mean test using F-test (with distinct mean) for Data 2

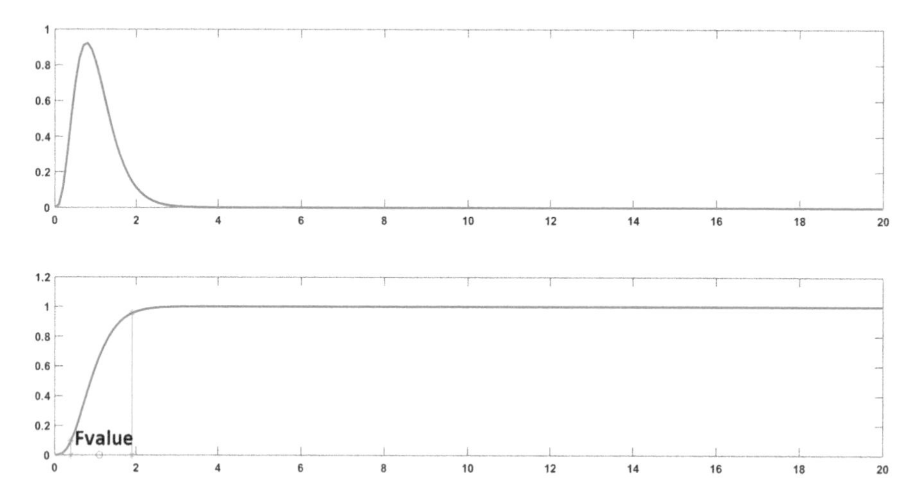

Fig. 6.8 Illustration to test for identical means of the individual groups using F-test (accepting the hypothesis that the means are identical with 90% confidence level)

```
  M{i}=mean(Group{i});
  end
%To check whether mean of the individual group are
identical
%using F-score
m=mean(cell2mat(M));
SB=sum(n.*((cell2mat(M)-repmat(m,1,c)).^2));
SW=0;
for i=1:1:c
```

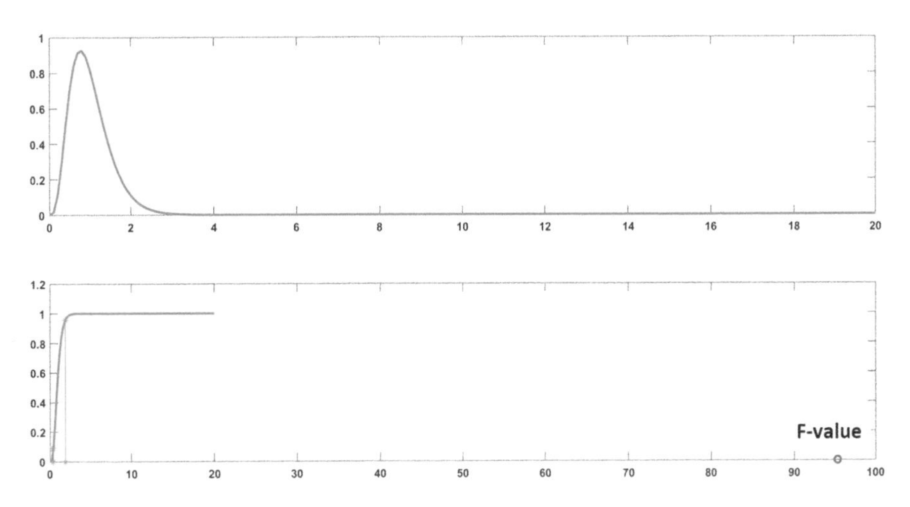

Fig. 6.9 Illustration to test for identical means of the individual groups using *F*-test (rejecting the hypothesis that the means are identical with 90% confident level) for data 2

```
    SW=SW+sum((Group{i}-repmat(M{i},1,length
    (Group{i}))).^2);
end
VARSW=SW/(sum(n)-c);
VARSB=SB/(c-1);
F=VARSB/VARSW;
%F follows F-distribution with order (c-1),(sum(n)-c)
Y=fpdf(0:0.1:20,c-1,sum(n)-c);
z=0:0.1:20;
Z=cumsum(Y)/10;
figure
subplot(2,1,1)
plot(z,Y,'r')
subplot(2,1,2)
plot(z,Z,'r')

%Locating 0.05 significant level (95% confident level)
%is the value at which the cumulative distribution is
%0.05 and 0.95
[p1,q1]=find((Z-0.05)>0);
[p2,q2]=find((Z-0.95)>0);
hold on
plot([z(q1(1)) z(q1(1))],[0 Z(q1(1))],'-r*')
hold on
plot([z(q2(1)) z(q2(1))],[0 Z(q2(1))],'-r*')
plot(F,0,'bo')
```

```
%The Null Hypotheis is structured such that the
%mean is equal to 90
%We reject the Hypotheis if F is greater than z(q1(1))
%or less than z(q2(1))
if((F<z(q1(1)))|(F>z(q2(1))))
    disp('Reject the Null Hypothesis (i.e., means are
    identical)')
else
    disp('Accept the Null Hypothesis (i.e., means are
    identical)')
end
figure
for i=1:1:c
plot(i*ones(1,n(i)),Group{i},'*')
hold on
end
```

6.7 Wilcoxon/Mann–Whitney Test (Median Test)

The mean test (refer Sects. 6.1–6.3) assumes that the data are from Gaussian distributed data. Median test can also be adopted to test whether median of the percentage of success obtained using the typical method is equal to hypothesized value. This is an alternative method for methods described in (6.1)–(6.3). The steps involved are summarized as follows:

1. Formulate the null hypothesis as median of the population is equal to m.
2. Let $X_1, X_2, X_3, \ldots, X_N$ be the outcome of the population.
3. $Y_i = X_i - m$ for all i.
4. Sort the absolute values of Y_i. Assign the ranks to the obtained data with 1st rank to the lowest value obtained. If the values are identical, reassign the ranks to those values as the average of the ranks assigned to them.
5. Assign sign to the obtained rank based on the actual difference.
6. Now perform the z-test/t-test identical to the method discussed in Sects. 6.1–6.3 (using the new data with signed rank values).

The illustration of Wilcoxon/Mann–Whitney test is illustrated in Fig. 6.10.

signedranktest.m

```
%Wilcoxon/Mann Whitney test (Signed rank test)
X=[82.9670   77.7751   90.6802   77.0560   79.2078
77.4048];
Y=X-80;
[P,Q]=sort(abs(Y));
```

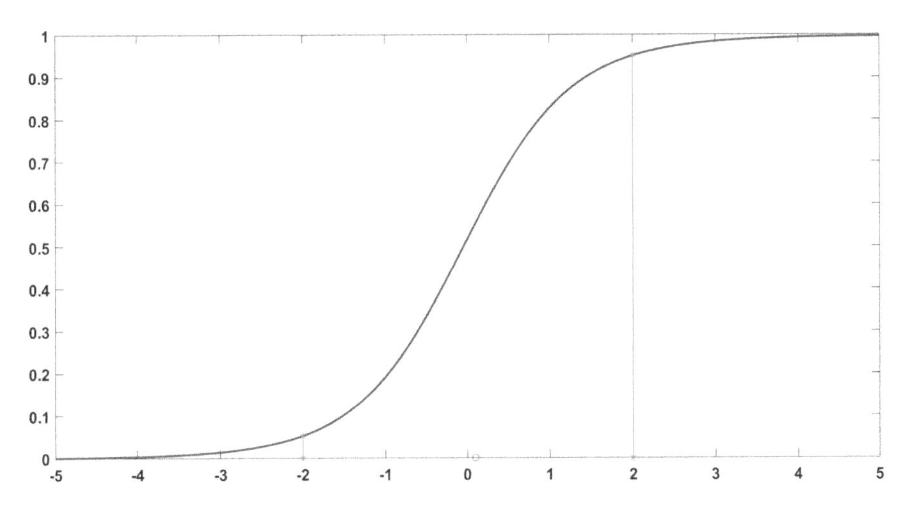

Fig. 6.10 Illustration of mean test using signed-rank method (Wilcoxon/Mann–Whitney test)

```
NEWDATA=Q.*sign(Y);
%Perform mean test using the modified data
m=mean(NEWDATA);
v=var(NEWDATA);
testt=m/sqrt(v/length(X));
%As the number of outcomes is lesser than $30$ and Z
%follows $t$ distribution with order 19
c=cumsum((pdf('t',-5:0.1:5,length(X)-1)/10));
t=-5:0.1:5;
plot(t,c);
%Locating 0.05 significant level (95% confident level)
%is the value at which the cumulative distribution is
%0.05 and 0.95
[p1,q1]=find((c-0.05)>0);
[p2,q2]=find((c-0.95)>0);
hold on
plot([t(q1(1)) t(q1(1))],[0 c(q1(1))],'-r*')
hold on
plot([t(q2(1)) t(q2(1))],[0 c(q2(1))],'-r*')
plot(testt,0,'bo')
%The Null Hypotheis is structured such that the
%mean is equal to 0 (Equivalently median of the
original data is equal to
%80)
%We reject the Hypotheis if ttest is greater than
t(q1(1))
%or less than t(q2(1))
```

```
if((ttest<t(q1(1)))|(ttest>t(q2(1))))
    disp('Reject the Null Hypothesis
(i.e., mean of the new data is equal to 0)')
else
    disp('Accept the Null Hypothesis
(i.e., mean of the new data is equal to 0)')
end
```

6.8 Kruskal–Wallis Test

As Wilcoxon/Mann–Whitney test is the alternative to mean test, Kruskal–Wallis test is the alternative to the F-test to get rid of outlier. The steps involved are as follows:

1. Arrange all the vectors in ascending order.
2. Assign ranks to the obtained data. Assign average rank values, if the data values are identical.
3. Compute S_W and S_B using the obtained rank values.
4. Follow the F-test as described in Sect. 6.6.

The illustration of Kruskal–Wallis test is given in Figs. 6.11, 6.13, 6.14, and 6.12.

kruskal-wallis-test.m

```
%Kruskal-Wallis test
c=10;
n=round(rand(1,c)*45)+5;
```

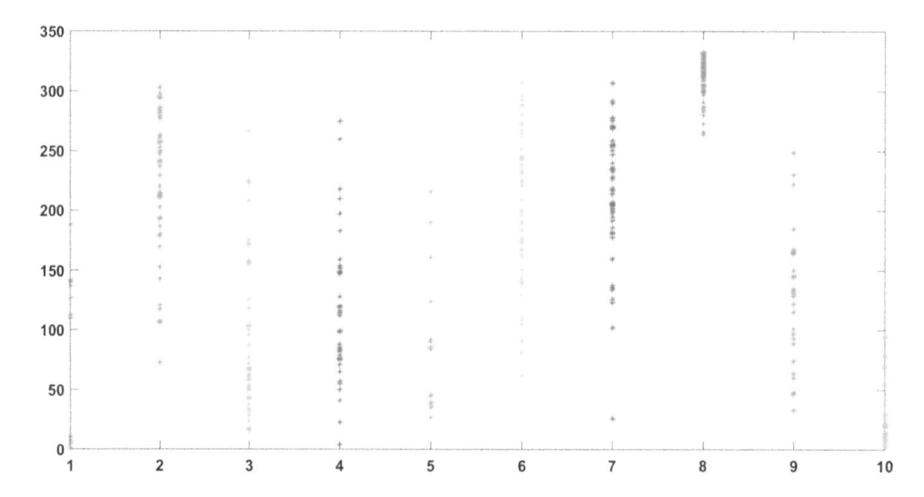

Fig. 6.11 Data 1 (with distinct mean) under mean test using Kruskal–Wallis F-test

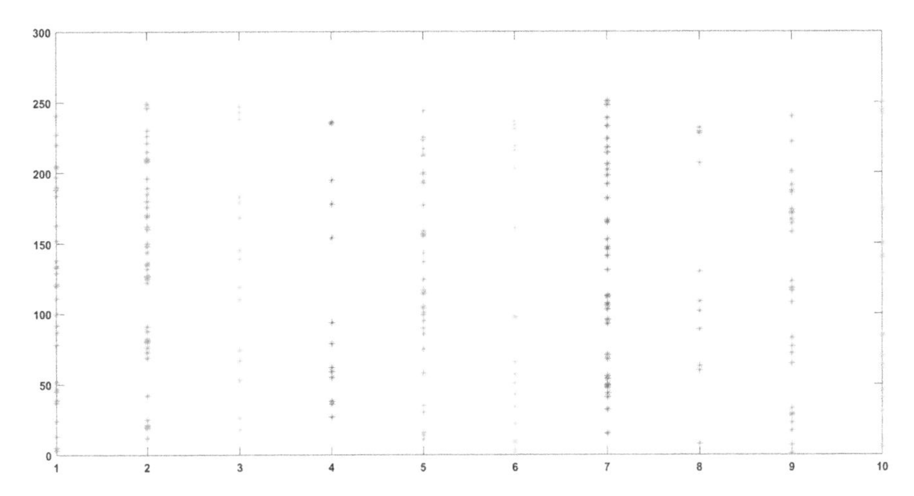

Fig. 6.12 Data 2 (with distinct mean) under mean test using Kruskal–Wallis F-test

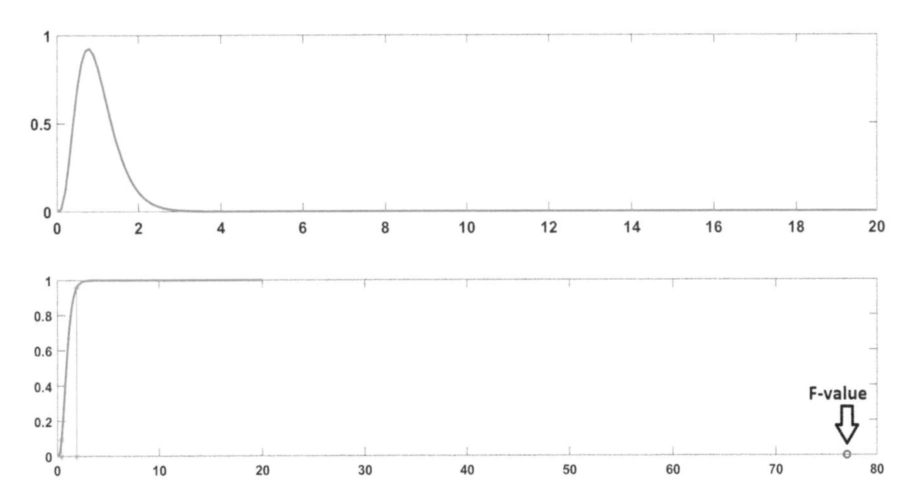

Fig. 6.13 F value obtained using Kruskal–Wallis test (data 1)

```
AM=round(rand(1,c)*39)+1;
for i=1:1:c
Group{i}=randn(1,n(i))*5+AM(i);
M{i}=mean(Group{i});
end
[P,Q]=sort(abs(cell2mat(Group)));
DATA(Q)=1:1:length(P);
Group=mat2cell(DATA,1,n);
for i=1:1:c
M{i}=mean(Group{i});
```

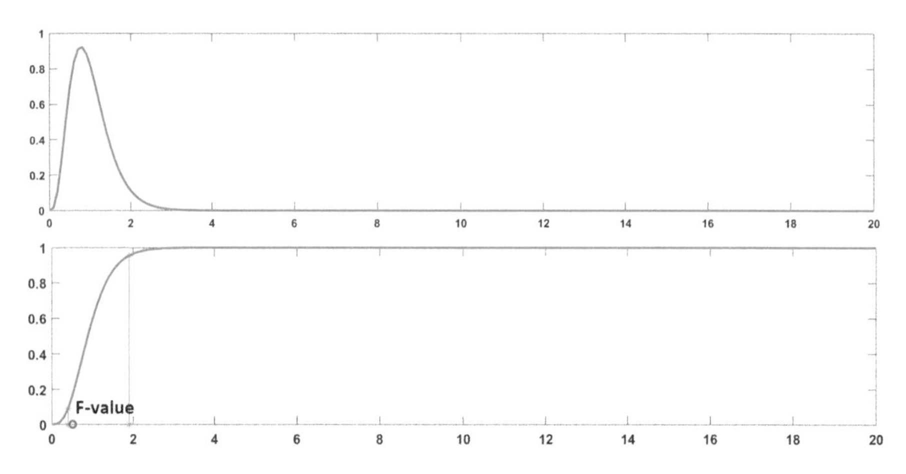

Fig. 6.14 *F* value obtained using Kruskal–Wallis test (data 2)

```
end
%To check whether mean of the individual group are
identical
%using F-score
m=mean(cell2mat(M));
SB=sum(n.*((cell2mat(M)-repmat(m,1,c)).^2));
SW=0;
for i=1:1:c
    SW=SW+sum((Group{i}-repmat(M{i},1,length
    (Group{i}))).^2);
end
VARSW=SW/(sum(n)-c);
VARSB=SB/(c-1);
F=VARSB/VARSW;
%F follows F-distribution with order (c-1),(sum(n)-c)
Y=fpdf(0:0.1:20,c-1,sum(n)-c);
z=0:0.1:20;
Z=cumsum(Y)/10;
figure
subplot(2,1,1)
plot(z,Y,'r')
subplot(2,1,2)
plot(z,Z,'r')

%Locating 0.05 significant level (95% confident level)
%is the value at which the cumulative distribution is
%0.05 and 0.95
[p1,q1]=find((Z-0.05)>0);
```

```
[p2,q2]=find((Z-0.95)>0);
hold on
plot([z(q1(1)) z(q1(1))],[0 Z(q1(1))],'-r*')
hold on
plot([z(q2(1)) z(q2(1))],[0 Z(q2(1))],'-r*')
plot(F,0,'bo')
%The Null Hypothesis is structured such that the
%mean is equal to 90
%We reject the Hypothesis if F is greater than
z(q1(1))
%or less than z(q2(1))
if((F<z(q1(1)))|(F>z(q2(1))))
    disp('Reject the Null Hypothesis (i.e., means are
    identical)')
else
    disp('Accept the Null Hypothesis (i.e., means are
    identical)')
end
figure
for i=1:1:c
plot(i*ones(1,n(i)),Group{i},'*')
hold on
end
```

List of m-Files

pca2d1d.m 2D to 1D conversion using PCA.

pcademo.m Demonstration of PCA.

fastpca.m Fast computation of PCA.

ldademo.m Demonstration of LDA.

kernellda.m Demonstration of kernel-LDA.

dimredlda.m Dimensionality reduction using LDA.

dimredpca.m Dimensionality reduction using PCA.

dimredklda.m Dimensionality reduction using KLDA.

visualizedata.m Visualization of data.

normalizedata.m Normalization of data.

genrandn.m Generate the outcome of the multivariate Gaussian density function.

gaussiankernel.m Gaussian kernel.

ICAdemo.m Demonstration of ICA.

kurt.m Kurtosis of data.

gradient.m Gradient of Objective function.

gaussianityafterKLDA.m Gaussianity applied to KLDA projected data.

maxgauss.m Maximizing Gaussianity of the given data.

discriminationdemo.m Demonstration on discrimination of multiple class problem.

© Springer Nature Switzerland AG 2020

E. S. Gopi, *Pattern Recognition and Computational Intelligence Techniques Using Matlab*, Transactions on Computational Science and Computational Intelligence,
https://doi.org/10.1007/978-3-030-22273-4

plotdata.m Used in discriminationdemo.m.

nearestmean.m Demonstration of nearest mean algorithm.

nearestneighbour.m Demonstration of nearest neighbour algorithm.

perceptron.m Demonstration of perceptron algorithm.

svmdemo.m Demonstration of Support Vector Machine (Hard margin).

innerproduct.m Inner product of the vectors.

softmargindemo.m Demonstration of Support Vector Machine (Soft margin).

softmargin1.m Used in softmargindemo.m.

softmargin2.m Used in softmargindemo.m.

svmforregression.m SVM for regression.

RVMforclassification.m Demonstration of Relevance Vector Machine.

regressiondemo.m Demonstration of linear regression.

Biasvardemo.m Bias variance decomposition demonstration.

performregression.m Used in Biasvardemo.m.

Bayesianapproachdemo.m Demonstration of Bayesian approach.

gbf.m Gaussian basis function.

gausskernelmethod.m Demonstration of Gaussian kernel method for regression.

regression-sequential.m Demonstration of sequential learning for linear regression.

Bayesregression.m Demonstration of Bayesian based regression.

predictivedistribution.m Demonstration of predictive distribution.

regression-RVM.m Demonstration of RVM for regression.

regression-RVMsparse.m Demonstration of sparsity based RVM for regression.

gaussianprocess.m Regression using Gaussian process.

crossentropy.m Demonstration of cross entropy.

logisticregression.m Demonstration of logistic regression for classification.

newtonraphson.m Newton–Raphson method.

logisticregression-newtonraphson.m Solving logistic regression using Newton–Raphson method.

GeneratedatausingGMM.m Generate the outcome of the Gaussian Mixture Model.

normalpdf.m Calculate the pdf of the multivariate Gaussian density function.

hmmgenerateseq.m Generate the binary sequence based on Hidden Markov Model.

genprob.m Calculate the generating probability, given the binary sequence generated from HMM.

sequencehmm.m Check for the sequence pattern based on the constructed HMM model.

generatehmmmodel.m Demonstration in obtaining HMM, given the binary sequence.

kmeans.m Demonstration of k-means algorithm.

psokurtosis.m Maximizing Gaussianity using kurtosis measurement using PSO.

objfun.m Objective function used in psokurtosis.m

antcolony-prob-detection.m Demonstration on obtaining the optimal order of the cascaded SVM blocks using ANT colony technique.

probcorrect.m Used in antcolony-prob-detection.m

seoakmeans.m k-Means algorithm using SEOA.

socialalgo.m Used in seoakmens.m.

SELO-sigma-KLDA.m Optimizing sigma of the Gaussian kernel used in KLDS using SELO.

seloalgo.m Used in SELO-sigma-KLDA.m

GApartition.m Genetic algorithm to obtain the partition line between two classes.

fga.m Used in GApartition.m.

BPNNtwoclass-softmax.m Backpropagation Neural Network using soft max activation function.

BPNN-multipletwoclass.m Backpropagation Neural Network using logsig activation function.

ANNregression.m ANN used as the regression technique.

BPNN-GMM.m ANN for Gaussian Mixture Model.

ccpdf.m Demonstration of Class conditional probability density function.

convolution-network.m Demonstration of Convolution Network.

GANdemo.m Demonstration of Generative Adversarial Network.

autoencoderdemo.m Demonstration of Autoencoder Network.

RNNdemo.m Demonstration of Recurrent Neural Network.

meantestwithknownvariance.m Statistical mean test with known variance.

meantest-withoutknownvariance.m Statistical mean test with unknown variance.

binomialtest.m Demonstration of Binomial test.

variancetest.m Demonstration of variance test.

proportiontest.m Demonstration of proportion test.

Fscore.m Demonstration of F-score.

signedranktest.m Demonstration of signed rank test.

kruskal-wallis-test.m Demonstration of Kruskal–Wallis test.

Index

© Springer Nature Switzerland AG 2020

E. S. Gopi, *Pattern Recognition and Computational Intelligence Techniques Using Matlab*, Transactions on Computational Science and Computational Intelligence,
https://doi.org/10.1007/978-3-030-22273-4

CPSIA information can be obtained
at www.ICGtesting.com
Printed in the USA
LVHW080433130620
657966LV00003B/18